Logic for Fun & Profit

Sanford W. Wood

Published by Indies United Publishing House, LLC

Cover Art by Danielle Johnston of irp.design

Book Design by Dave Reed

ISBN: 978-1-64456-037-2

Library of Congress Control Number: 2019905187

Indies United Publishing House, LLC
PO Box 3071
Quincy, Illinois 62305-3071
www.indiesunited.net

Contents

Preface

Let me explain the title. I have been teaching logic, off an on, for more than 40 years. I have found that I can make the class appear less rigorous and more fun by using a lot of entertaining examples. As for the "profit" part of the title, I explain to my students that there are many people (especially politicians and salesmen) who will use faulty logic in an attempt to persuade them to do things that they might be unwilling to do in more rational moments. So while the study of logic may not necessarily make you richer, it should help you keep the money you already have by enabling you to see through the sham arguments of con artists and snake-oil salesmen.

The style is light-hearted and witty, with humorous examples of logical errors. In addition, the text includes some local color reflecting some of the peculiarities of my Louisiana environment. Where rigor is required, clarity of exposition is supplemented with detailed examples. For instance, I show how the rules of the probability calculus can be derived from simple enumeration of the possibilities. Also, the rigor builds slowly, beginning with a warm-up that I call mental calisthenics, proceeding to humorous examples of misuse of language, and gradually moving to mathematical and computer logic.

As the table of contents indicates, this book is intended to be used for an introductory course in logic that would include both traditional and contemporary approaches to the study of logic. Based on my years of experience, I would say that the number of topics and their difficulty is just about right for a one-semester course. The text assumes no special background other than college readiness. In other words, it would not be an appropriate text for students requiring remediation in math or English.

The book is intended to be a stand-alone text requiring no special electronic technology. Each chapter contains sample exercises with solutions, followed by additional exercises that can be worked in class or assigned for students to work on their own. Answers to odd-numbered exercises can be found in the Appendix.

Introduction

Mental Calisthenics

Identifying Problems

In ordinary, everyday life, logic problems are not presented to us as they are in textbooks: neat, isolated, and identified as a particular type of problem. In everyday life, logic problems are tossed into ordinary conversation without any identifying labels. So our first challenge is being able to recognize a logic problem when it occurs. In the "aptitude test" that follows, some of the questions are trick questions, whereas others are real problems. The challenge, of course, is to decide which are which. If they are trick questions, explain the trick. If they are real problems, solve them.

1. Why is it illegal for a person living in Winston Salem, North Carolina, to be buried west of the Mississippi?
2. When I had eye surgery, I had to put four separate drops in my eye at 5-minute intervals. If I put the first drop in my eye at 8:00, when should I put the last one in?
3. Boudreaux has 25 alligators that he is raising for market. If he decides to slaughter all but 9 of them for their hides, how many will he have left?
4. Divide 30 by ½ and add 10. What is the answer?
5. A woman gives a beggar 50 cents. The woman is the beggar's sister, but the beggar is not the woman's brother. How come?
6. What is the one word that almost everyone pronounces wrong?
7. Is it legal for a man to marry his widow's sister?
8. A rope ladder is hanging over the side of a ship. The ladder is 12 feet long, and the rungs are one foot apart. The lowest rung is resting on the top of the ocean. The tide rises four

inches an hour. How long would it take before the first four rungs of the ladder are underwater?

9. Which would be worth more—a trunk full of nickels or the same size trunk half full of dimes?

10. In the Thibodeaux family there are seven sisters, and each sister has one brother. Including Mr. and Mrs. Thibodeaux, how many are in the family?

11. If the Vice-President of the U. S. should die in office, who would be the President?

12. Would it be cheaper for you to take one friend to the movies twice or two friends at the same time?

13. How close a relative would the sister-in-law of your father's only brother be?

14. According to the law of most states, the attempt to commit a certain crime is punishable, but committing the crime is not. What is the crime?

15. If you drop a steel ball, would it fall more rapidly through water at 20 degrees Fahrenheit or water at 60 degrees Fahrenheit?

16. The amount of water flowing into a tank doubles every minute. The tank is full in an hour. When was the tank half full?

Math Word Problems

Do you remember "word problems" from your high school math class? They were primarily logic problems, because the math involved in solving the problems is normally fairly trivial. Once you have identified how to set up the problem and which formula to use, the calculations are easy, especially if your teacher lets you use a calculator. These problems are like that. They look like math problems because they involve numbers, but they're really logic problems. You don't need to be a mathematician to solve them. You just need to think clearly.

1. There are 10 black socks and 10 white socks in a drawer. If you reach into the drawer in the dark, what is the least number of socks you must take out before you are sure of a pair that matches? Assume you do not care what color the socks are, as long as they match.

2. Suppose instead that there were 12 black socks and 8 white ones. How many socks do you need to pull out before you've got a match?

3. A man and his two sons must cross a stream. The man weighs 200 pounds and his sons each weigh 100 pounds. There is a boat that can carry only 200 pounds. How will they cross?

4. You have two containers; one holds five gallons, the other holds three. You can have as much water as you want. You are told to measure exactly four gallons of water into the five-gallon container. How do you do it?

5. In the traditional Christian calendar, there are two cycles of holy days, one based on Christmas, the other on Easter. Christmas has a fixed date, but the date of Easter is variable. It is defined as the Sunday after the first full moon of spring. Given this definition:

   - What is the earliest date on which Easter can occur?

   - What is the latest date?

6. Two trains are 500 miles apart, traveling toward each other on the same track. Train A is going 43 mph; Train B is going 57 mph. Meanwhile, a drone is flying back and forth between the two trains at 60 mph. How far will the drone fly (in total back-and-forth distance) before the two trains collide?

Puzzles

Puzzle-solving is a good mental warm-up for a logic class. The same kind of reasoning that is involved in solving a puzzle is also used in solving logic problems. The first step in solving a puzzle is to decide how the problem should be represented. If you can find a good way to represent the problem, you are well on your way to solving it. To understand the importance of representation, try to imagine how difficult it would be to do arithmetic using Roman numerals. Multiplying 25 by 13 is trivial, but try multiplying XXV by XIII (without converting).

Position Puzzles

In solving these puzzles, you need to represent them spatially. Once you have done so, the solution is fairly trivial. The first one is done for you as an example.

1. You are facing north. You turn to your right, do an about-face and turn to your right again. Which direction is at your back?

   Let's use arrows to represent direction, according to the traditional orientation of a map.

   You are facing north. ↑

   You turn to your right, →

   do an about-face ←

   and turn to your right again. ↑

   Which direction is at your back? By using the arrows, we see that you have turned full circle, so your back is to the south, as it was when you began.

2. The street on which I live, 4th St., is perpendicular to Maple St. Oak St. is parallel to Elm St. and perpendicular to 4th St. Is Maple perpendicular or parallel to Elm?

3. Three guys and three girls have been invited to a crawfish boil at Boudreaux's house. They all want to sit on the same side of one of Boudreaux's long tables (which seat six people on one side). No two people of the same gender want to sit next to each other. Tom does not want to sit by Jane. Dick is on one side of Jane and Harry is on the other side. April is in the first seat and Harry in the last one. Where does Carol sit?

4. Five friends live on the same street in a mobile-home park, and no one else lives on this street. The street has two double lots (that is, lots that will accommodate two mobile homes) and one single lot. The single lot is located between the two double lots. Tom does not share a lot with Dick, nor does he live next to him. Joe does not share a lot with Tom. You've been told that Harry and Moe either share a lot or live on adjacent lots. Who shares a lot with Joe?

Relationship Puzzles

There are different sorts of relationships involved in these puzzles. Some are comparisons among people based on factors such as height or age. Others are kinship relations. In each case we will use the information that is given in order to construct some sort of diagram that will represent the way the people are related to each other. The first one is done for you as an example.

1. John is taller than Bill. Tom is taller than Sam and shorter than Bill. Which one of these men is the tallest and which one is the shortest?

   First, we will make separate comparisons based on the two pieces of information. Differences in height are easily represented by putting the name of the taller person above the name of the shorter person.

   John is taller than Bill

   John
   Bill

   Tom is taller than Sam and shorter than Bill

   Bill
   Tom
   Sam

   Now we can combine the separate scales, and the answer is obvious:

   John
   Bill

Tom
Sam

2. John is younger than Bill. Tom is older than Bill. Sam is younger than John. Is Bill younger or older than Sam?

3. Three college coeds, April, Betty, and Carol, are majoring in literature, biology, and engineering, but not necessarily in that order. Betty took only introductory biology. Carol took just one math course because of a bad experience with math in high school, but she loves to read. April hates to read literature, and she is taking a math class with the engineer. What is each student's major?

4. Three college coeds, April, Betty, and Carol, each has a car. One is a Ford, one is a Chevy, and one is a Toyota. Betty owns a foreign car. Betty and April are in the same math class as the owner of the Ford. Match each car with its owner.

5. The contestants in the annual Bayou Teche swim meet were the alligator, the nutria, the duck, and the turtle. The nutria finished two places ahead of the duck. The alligator was neither first nor last and the nutria was not second. Who finished first and who finished last?

6. A, B, C, and D are all children. A is B's brother. B is C's sister. C is D's brother. From this information, what can we infer about the relationships and gender of these four children?

Matrix Puzzles

I call these matrix puzzles because it is useful to use a row-and-column or matrix format in representing the information that is given. This permits you to quite literally see how to solve the puzzle. The last one is done as an example.

1. Three United States senators – Sen. Jones, Sen. Smith, and Sen. Watson – find themselves on the same plane heading back to Washington after a brief Congressional recess. They represent the states of Alabama, Florida,

and Tennessee, though not necessarily in that order. By coincidence, three of the crew-members on the flight – the pilot, the co-pilot, and the navigator – have the same last names as the senators, and live in the same three states, though not necessarily in that order. We have this additional information:

- Sen. Watson lives in Alabama.
- Even though Sen. Smith is something of an artist, he knows nothing about the greeting cards business.
- Crewman Jones won the last three gin rummy games with the co-pilot.
- The passenger whose last name is the same as the navigator's lives in Tennessee.
- The navigator lives in Florida.
- The navigator's next-door neighbor, who is one of the senators, has been in the greeting-card business for 20 years.

Where do each of the senators live? What are the positions of each of the members of the flight crew?

2. The finalists in the National Spelling Bee are Carol, Dick, Francis, George, and Myrtle. They come from various states – Colorado, Delaware, Florida, Georgia, and Maryland – but not in that order. They have stated their intended college majors as computer science, drama, finance, geography, and medicine, but not in that order. In fact, none of the contestants comes from a state beginning with the same letter as his or her name, and their intended major begins with a different letter than either their name or the state from which they come. We know these additional facts:

    - The drama major does not come from Florida.
    - Dick is not planning to major in either computer science or medicine, and he is not from Florida or Georgia.

- George is from Maryland, but he is not majoring in computer science or drama.
- Neither Carol nor Myrtle is from Delaware.
- Myrtle is not majoring in drama or geography.

Determine the state represented by each of these contestants and their intended college majors.

3. The employees of a small loan company are Mr. Black, Mr. White, Mrs. Coffee, Miss Ambrose, Mr. Kelly, and Miss Earnshaw. The positions they occupy are manager, assistant manager, cashier, stenographer, teller, and clerk, though not necessarily in that order. The assistant manager is the manager's grandson, the cashier is the stenographer's son-in-law, Mr. Black is a bachelor, Mr. White is twenty-two years old, Miss Ambrose is the teller's step-sister, and Mr. Kelly is the manager's neighbor. Who holds each position? (Note: The solution of this puzzle is my own, but the puzzle came from another logic book. I would like to acknowledge the author, but unfortunately I have lost the reference.)

Below is the matrix before any information is recorded. Column headings represent the employees and row headings represent the positions.

	Miss A	Mr. B	Mrs. C	Miss E	Mr. K	Mr. W
Manager						
Asst. Mgr.						
Cashier						
Stenog.						
Teller						
Clerk						

To solve this puzzle, we have to assume that everyone is "proper", that is, no children out of wedlock, no incestuous relationships, etc. Let's take it one bit of information at a time, and gradually fill in the matrix.

The assistant manager is the manager's grandson.

- This means that the assistant manager is male, so we can eliminate all the women.
- Also, the manager must be married (or have been), so we can eliminate all the single women.

	Miss A	Mr. B	Mrs. C	Miss E	Mr. K	Mr. W
Manager	X			X		
Asst. Mgr.	X		X	X		
Cashier						
Stenog.						
Teller						
Clerk						

The cashier is the stenographer's son-in-law.

- The cashier must be male, so we can eliminate all the women.
- Also, the stenographer must be married (or have been), so we can eliminate all the single women.

	Miss A	Mr. B	Mrs. C	Miss E	Mr. K	Mr. W
Manager	X			X		
Asst. Mgr.	X		X	X		
Cashier	X		X	X		
Stenog.	X			X		
Teller						
Clerk						

Mr. Black is a bachelor.

- He cannot be the manager or stenographer, because they are married.
- He cannot be anyone's son-in-law, so he is not the cashier.

	Miss A	Mr. B	Mrs. C	Miss E	Mr. K	Mr. W
Manager	X	X		X		
Asst. Mgr.	X		X	X		
Cashier	X	X	X	X		
Stenog.	X	X		X		
Teller						
Clerk						

Mr. White is twenty-two years old.

- He's not old enough to be a grandfather, so he cannot be the manager.
- He's not old enough to have a son-in-law, so he cannot be the stenographer.

	Miss A	Mr. B	Mrs. C	Miss E	Mr. K	Mr. W
Manager	X	X		X		X
Asst. Mgr.	X		X	X		
Cashier	X	X	X	X		
Stenog.	X	X		X		X
Teller						
Clerk						

Miss Ambrose is the teller's step-sister.

- Miss Ambrose is not the teller.
- Miss Ambrose must be the clerk (because there is no other position left for her).
- If Miss Ambrose is the clerk, then we can X out "Clerk" for everyone else.
- Thus, Miss Earnshaw must be the teller (because there is no other position left for her).
- If Miss Earnshaw is the teller, then we can X out "Teller" for everyone else.

- Thus, Mr. Black must be the assistant manager (because there is no other position left for him).
- If Mr. Black is the assistant manager, then we can X out "Asst. Mgr." for everyone else.
- Thus, Mr. White must be the cashier (because there is no other position left for him).
- If Mr. White is the cashier, then Mr. Kelly cannot be.

	Miss A	Mr. B	Mrs. C	Miss E	Mr. K	Mr. W
Manager	X	X		X		X
Asst. Mgr.	X	***	X	X	X	X
Cashier	X	X	X	X	X	***
Stenog.	X	X		X		X
Teller	X	X	X	***	X	X
Clerk	***	X	X	X	X	X

Mr. Kelly is the manager's neighbor.

- Mr. Kelly cannot be the manager.
- Thus, Mr. Kelly must be the stenographer (because there is no other position left for him).
- Thus, Mrs. Coffee must be the manager (because there is no other position left for her).

	Miss A	Mr. B	Mrs. C	Miss E	Mr. K	Mr. W
Manager	X	X	***	X	X	X
Asst. Mgr.	X	***	X	X	X	X
Cashier	X	X	X	X	X	***
Stenog.	X	X	X	X	***	X
Teller	X	X	X	***	X	X
Clerk	***	X	X	X	X	X

Logic and Language

The Importance of Language

Language is a marvelous thing. It can be both delightful and annoying. You can use it to make friends or to make enemies, to lift people's spirits or to dash their hopes, to dazzle them with your brilliance or to expose yourself as a pompous fool.

Why language is so important to the study of logic:

- Even though it is possible to use logic without using language, most people cannot do so.
- The most important misuses of logic occur in persuasive uses of language. We need to be on our guard.
- A recurring problem is how to "translate" from ordinary English into a logical standard form that will permit analysis.

Uses of Language

Any attempt to classify the ways in which language is used will invariably be incomplete and arbitrary. What follows is an attempt to provide some idea of the range of language uses.

Communication with other people:

- Exchange of information: "Walmart is having a big sale."
- Instruction: "Before you jack up your car, be sure to block the wheel on the opposite side."
- Translation: "When a kid says something is bad, he really means it's good."
- Giving orders: "Lift that barge, tote that bale."
- Making threats: "If you don't eat your vegetables, you won't get any dessert."
- Persuasion/seduction: examples range from a carefully reasoned argument in a scholarly journal to a used-car salesman's pitch.

- Questioning: "Have you read *King Lear*?"

Non-communicative public uses: Note that information may be communicated, but that is not the language's primary function.

- Greetings: "Hi, there."
- Small talk: "It's a hot one today."
- Requests: "Please pass the salt."
- Ceremonial: ". . . in the name of the Father, and of the Son, and of the Holy Spirit."
- Official: "We find the defendant guilty as charged."
- Beautiful (artistic format): poetry, drama, rhetoric. Note that bad poetry (etc.) also fits into this category.
- Impressive: as when people use fancy language when plain language would do just as well, such as Dr. Johnson's definition of a net as a "fabric with interstitial vacuities". Or "Fortunately, problems are easily assuaged in all but the most intractable instances."
- Humor: "My cow just died so I don't need your bull."
- Inspirational/motivational: evangelism, pep talks, etc.

Private uses:

- Expressions of joy, anger, etc: "Damn!" after you hit your thumb with a hammer.
- Recording information in a diary, making notes to yourself, etc. Note: usually written, not spoken.
- Thinking in words.

Language Forms and Language Functions

Language can take various forms. As before, there is no claim that this list is complete.

Ideally, the form of a statement provides a clear indication of the statement's function, as in the following examples.

- Declarative: "The square root of 4 is 2."

- Interrogative: "Can you recommend a good plumber?"
- Exclamatory: "I won the lottery! I won the lottery!"
- Performatory: "I pronounce you man and wife."
- Poetic: use of rhyme and/or rhythm: "Jack and Jill went up the hill to fetch a pail of water . . ."

Sometimes form misleads, as in these examples:

- "Do you know how to get to Chicot State Park?" The technically correct answer is Yes, but that's probably not what is wanted. It is a request for information.
- "People who ask a lot of questions do not last long in this organization." It has the form of a declarative statement, but it is actually a threat.

In each of the following passages, identify the form of the statement or statements and the primary language function that the passage is probably intended to serve. Explain your answer. The first one is done for you as an example.

1. A passage from the book of Wisdom that is customarily read during the service for All Saints Day: "Let us now sing the praises of famous men, our ancestors in their generations. Their bodies are buried in peace, but their name lives on generation after generation."

   These sentences are declarative statements. When they were originally written, they were probably intended to be inspirational or motivational. In their present use, as part of a religious ceremony, their function is primarily ceremonial.

2. Mayor speaking to honored guest: "I hereby declare you an honorary citizen of our fair city."

3. From the Nicene Creed: "I believe in God the father almighty, maker of heaven and earth."

4. From the Pythagorean Theorem: "The square on the hypotenuse is equal to the sum of the squares on the other two sides."

5. From an explanation of the theoretical foundations of computer science: "Every actual computer is an implementation of a mathematical model known as a Turing machine."

6. From the Sermon on the Mount: "Blessed are the peacemakers, for they shall be called children of God."

7. From the Rubaiyat of Omar Khayyam:
   "A book of verses underneath the bough,
   A loaf of bread, a jug of wine, and thou
   Beside me, singing in the wilderness;
   Ah! wilderness were paradise enow."
   (Note: "enow" is an older form of "enough".)

8. From one of Dave Barry's columns: "Karate is a form of martial arts in which people who have had years and years of training can, using only their hands and feet, make some of the worst movies in the history of the world."

9. From Darwin's Descent of Man: "Man scans with scrupulous care the character and pedigree of his horses, cattle, and dogs before he mates them; but when he comes to his own marriage he rarely takes any such care."

10. Slogan on a T-shirt: "Fighting for peace is like making love for virginity."

11. The technical version of what is popularly known as the theory of gravity: "Every physical object attracts every other physical object, directly according to the product of their masses, and inversely according to the square of the distance between them."

12. A statement attributed to Abraham Lincoln: "It's better to remain silent and be thought a fool than to speak and remove all doubt."

13. From the Epistle to the Hebrews: "Faith is the substance of things hoped for, the evidence of things not seen."

14. Slogan on a T-shirt: "Hard work has a future payoff. Laziness pays off now."

Misuses of Language

To parody Browning, we can say to language, "How can I misuse thee? Let me count the ways." Whenever I think I have seen all the ways that language can be misused, my students prove me wrong, so I am confident this list is not complete.

Syntax problems:

- Grammatical errors: a campaign poster for a legislator running for re-election: "Nobody does nothing better than Jones." Is it a grammatical error or truth in advertising?
- Mixed metaphor: "Take the bull by the hand."
- Fragmentary or run-on sentences: "Don't write a run-on sentence you got to punctuate it."
- Misplaced modifier: "Wrapped only in a newspaper, the lady carried the fish down the street."
- Unintended double meaning: "Why go elsewhere to be cheated? Come here first."

Logic problems:

- Inaccuracy in the use of words: "I was literally climbing the walls."
- Ambiguity: using a word or phrase in two different ways: "I still miss my ex-wife, but my aim is getting better."
- Vagueness: "There weren't many people in church today." How many? This is relative to normal attendance. In the National Cathedral, this statement would be accurate if only 100 people showed up; but 100 people would be "many" in the small church that I attend.
- Irrelevance: when the prosecution has a weak case, they may go to great lengths to prove how horrible the crime was -- all of which is irrelevant to the question of whether the accused is guilty.
- Nonsense: "The square root of butter is love."

Examples of Mixed Metaphors

- The sacred cows have come home to roost.
- She'll get it by hook or ladder.
- These hemorrhoids are a real pain in the neck.
- That snake in the grass is barking up the wrong tree.
- It's time to swallow the bullet.
- Let dead dogs sleep.
- The guy's out to butter his own nest.
- She bends overboard to be fair.
- He's between a rock and the deep blue sea.

Examples of Misplaced Modifiers

- Bound, gagged, and trussed up nude in a denim bag, the teacher explained how she was kidnapped.
- Plunging 1000 feet into the gorge, we saw Yosemite Falls.
- Calf born to farmer with two heads!
- He rode his horse up and down the sidewalk in front of the bar for a good half hour before deputies arrived, shouting obscenities and being obnoxious.
- Breaking into the window of the girls' dorm, the dean of men surprised 10 members of the football team.
- The patient was referred to a psychiatrist with a severe emotional problem.
- A 30-year-old man was found murdered by his parents in his home late Saturday night.
- The judge sentenced the killer to die in the electric chair for the second time.
- No one was injured in the blast, which was attributed to a build-up of gas by one town official.

Examples of Unintended Double Meaning

1. This is something I got off the Internet years ago. It was

claimed that these sentences actually appeared in church bulletins or were announced in a church service:

- Bertha Belch, a missionary from Africa, will be speaking tonight at Calvary Memorial Church in Racine. Come tonight and hear Bertha Belch all the way from Africa.
- Miss Charlene Mason sang "I will not pass this way again", giving obvious pleasure to the congregation.
- Ladies, don't forget the rummage sale. It's a chance to get rid of those things not worth keeping around the house. Don't forget your husbands.
- The sermon this morning: "Jesus Walks on the Water". The sermon tonight "Searching for Jesus".
- The Rector will preach his farewell message after which the choir will sing "Break Forth into Joy."
- Don't let worry kill you off. Let the Church help.
- Irving Benson and Jessie Carter were married on October 24 in the church. So ends a friendship that began in their school days.
- A bean supper will be held on Tuesday evening in the church hall. Music will follow.
- At the evening service tonight, the sermon topic will be "What is Hell?" Come early and listen to our choir practice.
- Eight new choir robes are currently needed, due to the addition of several new members and to the deterioration of some older ones.
- For those of you who have children and don't know it, we have a nursery downstairs.
- Please place your donation in the envelope along with the deceased person(s) you want remembered.
- Ladies Bible Study will be held Thursday morning at 10. All ladies are invited to lunch in the Fellowship Hall after the BS is done.

- Low Self Esteem Support Group will meet Thursday at 7 PM. Please use the back door.
- The eighth-graders will be presenting Shakespeare's Hamlet in the Church basement Friday at 7 PM. The Congregation is invited to attend this tragedy.
- Weight Watchers will meet at 7 PM at the First Presbyterian Church. Please use large double door at the side entrance.
- The Associate Minister unveiled the church's new tithing campaign slogan last Sunday: "I Upped My Pledge - Up Yours."

2. Advertisements can also be an entertaining source of unintended double meaning.
   - For sale: antique desk suitable for lady with thick legs and large drawers.
   - Four-poster bed, 101 years old, perfect for antique lover.
   - Dog for sale. Eats anything and is fond of children.
   - Man wanted to work in dynamite factory. Must be willing to travel.
   - Wanted: man to take care of cow that does not smoke or drink.
   - Our bikinis are exciting. They are simply the tops.
   - Mixing bowl designed to please a cook with round bottom for efficient beating.

3. The following headlines are reputed to be authentic. Since I got them over the Internet, the claim must surely be valid.
   - Police Begin Campaign to Run Down Jaywalkers
   - Panda Mating Fails; Veterinarian Takes Over
   - Clinton Wins on Budget, But More Lies Ahead
   - Enraged Cow Injures Farmer with Ax
   - Miners Refuse to Work after Death

- Juvenile Court to Try Shooting Defendant
- Stolen Painting Found by Tree
- Two Sisters Reunited after 18 Years in Checkout Lane
- Killer Sentenced to Die for Second Time in 10 Years
- Red Tape Holds Up New Bridge
- New Study of Obesity Looks for Larger Test Group
- Kids Make Nutritious Snacks

Malapropisms

Malapropisms constitute a special category of inaccuracy in the use of words. A malapropism is a word or phrase that sounds like another word or phrase that has a very different meaning. For example, the words *prostate* and *prostrate* sound very much alike, but their meanings are far apart. A malapropism occurs when the wrong one is used, as in the following examples:

- He lay prostate upon the floor.
- Poor Joe has been diagnosed with prostrate cancer.

Former President George W. Bush was notorious for creative misuse of the English language. This is one of his most famous malapropisms:

> We cannot let terrorists and rogue nations hold this nation hostile or hold our allies hostile.

It is clear from the context that he meant to say *hostage* instead of *hostile.*

Malapropisms frequently occur when people are trying to use words that are beyond their ordinary vocabulary. They sometimes occur in writing when a person uses a word or phrase that sounds right, but has a very different meaning. My favorite example is from an ethics class. I had asked the students to write a brief essay on euthanasia. One girl turned in an essay entitled Youth in Asia.

Some humorous examples of malapropisms occur in children's misunderstanding of adult conversation. I recall reading about one little girl who had visited a friend's church. When she got home, she told her mother that they sang a song about a bear named Gladys who didn't see well. When her mother inquired, she found out that the name of the hymn was *Gladly the Cross I Bear*, but the little girl had understood it as *Gladys the Cross-Eyed Bear.*

Sometimes humorists will deliberately create examples of malapropisms based on the hearing difficulties associated with old age. I recall seeing a cartoon showing three old men sitting on a park bench. The first one said, "It sure is windy today." The second one replied, "No, it's not Wednesday; it's Thursday," to which the last one responded, "I'm thirsty, too. Let's grab a beer."

In the following examples, the intended word or phrase appears in parentheses at the end of the statement.

1. Rainy weather can be hard on the sciences. (sinuses)
2. Alice said she couldn't eat crabs or any other crushed Asians. (crustaceans)
3. His speech contained several literary delusions. (allusions)
4. You lead the way and we will precede. (proceed)
5. A rolling stone gathers no moths. (moss)
6. Having only one wife is called monotony. (monogamy)
7. The flooding was so bad they had to evaporate the city. (evacuate)
8. John has a pornographic memory. (photographic)
9. Flying saucers are just an optical conclusion. (illusion)
10. I remember going to a menstrual show when I was a kid. (minstrel)
11. As a woman gets older, she often develops very close veins. (varicose)
12. Pres. George W. Bush is alleged to have said, 'If it wasn't for the electrical college, I wouldn't be president.'" (electoral)

13. The minister's five-year-old son and his playmates had found a dead robin. Feeling that proper burial should be performed, they had secured a small box and cotton batting, then dug a hole and made ready for the disposal of the deceased. The minister's son was chosen to say the appropriate prayers and with sonorous dignity intoned his version of what he thought his father always said. "Glory be unto the Faaaather...and unto the Sonnnn...and into the hole you goooo." (unto the Holy Ghost)

Ambiguity and Vagueness

The distinction between ambiguity and vagueness is not as clear and precise as we might like it to be. Some logic texts classify statements as vague that I would consider to be ambiguous. My understanding of ambiguity is that it refers to statements in which an expression has two clearly defined meanings, but the context makes it unclear which meaning is intended. By contrast, vagueness refers to statements that have no precise or clearly defined meaning.

Ambiguity

There are two generally accepted categories of ambiguity:

- Equivocation refers to ambiguity caused by a term that has two well-known meanings. Often there is a change of meaning within the statement, as in my earlier example: "I still miss my ex-wife, but my aim is getting better."
- Amphiboly (or amphibology) designates ambiguity caused by an awkward grammatical structure. Consider the statement: "The teachers asked the students to read the papers they had written." Who had written the papers, the teachers or the students? Either meaning is possible, and we can't tell which meaning was intended. (Note: This is an example of a statement that is classified as vague by some logicians.)

In each of the following examples, explain whether the ambiguity is caused by an ambiguous word or an awkward grammatical structure. You should also specify the two meanings. The first one is done for you as an example.

1. Only man creates art. No woman is a man. Therefore no woman creates art.

   Equivocation on the word "man". In the first sentence, "man" is used to refer to a species; in the second sentence, it designates a gender.

2. We were fighting for freedom, but freedom is just a word. Fighting for words is stupid.

3. If a tree falls in the forest and there's no one there to hear it, then it doesn't make a sound.

4. You thought my yacht was bigger than it is, but only an idiot would think that something is bigger than it is.

5. After Maggie insulted Mary she slapped her.

6. Boudreaux: You say that when I get to the third traffic light, I turn left?
   Dugas: Right.
   Boudreaux: But you just said to turn left.

Vagueness

There is, to my knowledge, no standard way of classifying vague statements. However, the following two categories are fairly common:

- Quantitative vagueness is caused by the use of quantitative terms that have no precise meaning – words such as "many", "few", "some", etc. Refer to my earlier example: "There weren't many people in church today."
- Referential vagueness is caused by statements in which the frame of reference is not specified. Much statistical information is presented in this way. "There were three fatal accidents over the holiday weekend." What is the frame of reference? If we're speaking about the whole country, this is insignificant; but three fatalities in the small town in which I live would be noteworthy.

In the following examples, identify the word or phrase that is causing the vagueness. Also specify what sort of vagueness it is. The first one is done for you as an example.

1. From an antacid commercial: "Faster, longer lasting relief."

   Faster and longer lasting than what? Than our competitor's product? Than what we sold you last time? This sort of vagueness is typical of commercial messages, which aim at creating a good feeling without providing specific information. This could be called quantitative vagueness because it doesn't specify how much faster and how much longer lasting, but it could also be called referential vagueness because we don't know what the product is being compared to.

2. Walmart: always the low price.

3. Government warning: "Excessive exposure to the sun causes skin cancer."

4. This car gets excellent mileage. They say it goes 25 percent farther on a tank of gas.

5. From a whiskey ad: "Impress friends and influence people. Seagram's V.O. Bottled in Canada. Preferred throughout the world."

6. Be sure you are well-dressed for your interview.

Analogies

Analogies are generally used to make a point. The "point" can be made in support of some position or in opposition to it. Let's look at examples of each.

Support: "Logic is like playing the piano: the more you practice, the more proficient you become." This is a positive analogy. It is probably intended to encourage struggling logic students by reminding them that no one becomes a good pianist overnight, so they should not become discouraged if they don't understand logic very well at first.

Opposition: "Christian Science is like grape nuts, neither the one nor the other." This is a "put-down" analogy. Whoever made the comparison doesn't think much of Christian Science. He (she) is saying in effect that Christian Scientists are neither Christians nor scientists, just as grape nuts are neither grapes nor nuts.

Sometimes analogies are implied rather than stated explicitly. Consider the following statement by the philosopher George Santayana:

> The philosophy of the common man is an old wife who gives him no pleasure, yet he cannot live without her, and resents any aspersions that strangers may cast on her character.

Santayana does not use any terms of comparison, yet it is obvious that he does not intend his statement to be taken literally. He does not mean that the philosophy of the common man *is* an old wife, but that it is *like* an old wife.

There is another category of analogy, namely, those that are created as jokes. This is one of my favorite analogies: "A herd of buffalo can only move as fast as the slowest buffalo. And when the herd is hunted, it is the slowest and weakest ones at the back that are killed first. This natural selection is good for the herd as a whole, because the general speed and health of the whole group keeps improving by the regular killing of the weakest members. In much the same way, the human brain can only operate as fast as the slowest brain cells. Excessive intake of alcohol, we all know, kills brain cells, but naturally it attacks the slowest and weakest brain cells first. In this way, regular consumption of beer eliminates the weaker brain cells, making the brain a faster and more efficient machine. And that's why you always feel smarter after a few beers."

In evaluating analogies, we should remember that an analogy is never exact. It involves a comparison of two things that are alike in some respects, different in others. For an analogy to be effective, the similarities should outweigh the differences. While there are no precise rules for evaluating analogies, we can give some general guidelines.

1. An analogy should be <u>more familiar</u> than the topic that is at issue. For example, it is not a good analogy to say that something is "as counter-intuitive as Reimonian

geometry", because most people have not heard of Reimonian geometry.

2. An analogy should be enlightening, that is, it should provide new information or a fresh perspective. A good example would be: "Men are like big children. Their toys are just more expensive."

3. An analogy should not prejudice the issue by making a comparison that is obviously biased. For example, some opponents of abortion compare abortion to the primitive practice of infanticide. Since this appears to be a biased comparison, it is not likely to be effective except with people who already believe that abortion is wrong.

4. The similarities between the things being compared should be more important than their differences. For example, if someone says that cigarettes are like coffin nails, the analogy is effective only if we don't look at it too closely. Coffin nails are associated with death, but they don't cause it.

In evaluating the following analogies, try first to determine the point that is being made, then say whether the analogy is effective in making that point. The first one is done for you as an example.

1. A woman needs a man like a fish needs a bicycle.

   This appears to be a put-down. If its purpose is to show how useless men are, it's pretty good. (Are we really that bad?)

2. An argument by analogy is like a rented tuxedo: it never quite fits.

3. Comparing a baseball team to a basketball team is like comparing apples to oranges.

4. Karl Marx said that anyone who was a salaried employee was a "wage slave."

5. Because they're usually the center of attention in a room, television sets function much the same way that fireplaces did a hundred years ago.

6. War is hell.

7. The eyes are the windows of the soul.
8. The president of the United States is the head of the nation.
9. Business is a rat race.
10. The brain secretes thought as the stomach secretes gastric juice, the liver bile, and the kidneys urine.
11. A compassionate conservative is like a vegetarian cannibal.
12. A female perspective: Men are like fine wine. They all start out like grapes, and it's our job to stomp on them and keep them in the dark until they mature into something with which you would like to have dinner.
13. A male perspective: Women are like fine wine. They all start out fresh, fruity, and intoxicating to the mind. They turn full-bodied with age, then they go all sour and vinegary and give you a headache.
14. As in prospecting for gold, a scientist may dig with skill and intelligence just a few feet away from a rich vein without success. Consequently, in scientific research the rewards for industry and intelligence are highly uncertain.
15. Einstein is said to have criticized Heisenberg's uncertainty principle by saying, "God does not play dice with nature."
16. Perhaps the most startling discovery made in astronomy this century is that the galaxies in the universe are systematically receding from one another, like raisins in an expanding pudding.

Traditional Logic

Arguments

The Terminology of Logic

Up to this point, I have been talking a good bit about logic without having first defined it. It's as if I am assuming that everybody knows what logic is. The problem with that assumption is that what "everybody knows" is often quite variable. Years ago I operated a hot dog stand. I soon discovered that "everybody knows" what should be put on a hot dog, but almost everyone knew something different. Some people thought that a "regular" hot dog was just a wiener in a bun. Some thought it should have chili on it, others said mustard and onions, and so on. So maybe I should attempt to define logic before going any further.

By way of a preliminary definition, we can say that logic is the study of reasoning. "But what kind of study?" you may ask. Is it descriptive or prescriptive? Do we study the way people actually reason, or the way they should? Well, we will do a bit of both. When we discuss fallacies, we will be examining recurring patterns of incorrect reasoning. This is the way many people actually do reason. We need to be able to identify these errors when they occur in other people's reasoning, and avoid them in our own. We will also study rules and procedures for correct reasoning.

Of course, reasoning is a very large topic, so we need to restrict our scope somewhat. Logic is primarily concerned with a special type of reasoning called *inference*: the claim that *because* we know X and Y are true, *therefore* Z must also be true.

Now a bit of terminology:

- What we are trying to prove is called the *conclusion.*
- The reasons that we give for the conclusion are called *premises* (one or more).
- Premises and conclusion together are called an *argument.*
- The process of reasoning from premises to conclusion is called *inference.*

There's another bit of terminology that requires a rather lengthy explanation. I'm referring to the distinction between *statements* and *propositions*. A statement has a specific form, for example, "Truman beat Dewey in the Presidential election of 1948." But I could just as well have said, "Dewey lost to Truman in the Presidential election of 1948." These two statements have the same meaning. It is the meaning that logicians refer to when they use the word "proposition". Instead of saying that two statements mean the same thing, logicians would say that they express the same proposition. The significance of this for the study of logic is that we can change the structure and wording of a statement any way we wish, as long as we do not change its meaning.

Having said that, I must hastily add that logic is only concerned with a small portion of what we normally have in mind when we talk about the meaning of something. Specifically, logic is concerned with truth-value – whether a statement is true or false. But there are many nuances and subtleties, connotations and suggestions, that are not part of what logicians call "meaning". Consider the following passage: "Oh for a draft of vintage that hath cooled a long age in the deep-delved earth!" Now, what does this mean? A very prosaic rendering might go something like this: "I would like to have some fermented grape juice whose temperature has been reduced because it was left in a deep hole for a long time." Something seems to have been lost in translation. The truth-value is captured, but very little else.

In the following exercises, some anonymous Internet author has done with Christmas carols what I just did with the passage from Keats: they have rewritten the titles in an approximate prosaic equivalent. Your task is to identify the carol. The first two are done for you as examples.

1. My sole desire for the Yuletide season is receipt of a pair of central incisors.

   All I want for Christmas are my two front teeth.

2. From dark 'til dawn, soundless and sanctified.

   Silent night, holy night.

3. Celestial messengers from splendid empires.

4. The antlered quadruped with the cerise proboscis.
5. The event occurred at one minute after 11:59 p.m. with visibility unlimited.
6. Ornament the enclosure with large sprigs of berry-bearing evergreen.
7. Personal hallucinations of an alabaster December 25th.
8. Testimony of witness to maternal parent's infidelity with Kris Kringle.
9. In a distant bovine diner.
10. Universal elation.
11. O miniature Nazarene village.
12. The approach of the holiday commemorating the birth of Christ is becoming evident.
13. Those of you who are true, come here!
14. Are you detecting the same aural sensations as I am?
15. The diminutive male of less than adult age who plays a percussion instrument.
16. Primary Yuletide song.
17. Heavenly cherubim announcing in song – Listen!

Another distinction that requires some elaboration is the distinction between *arguments* and *disagreements*. In ordinary language these terms tend to be used somewhat interchangeably, but logicians have given the term "argument" a technical meaning (as indicated above). Following their lead, we will use the term "argument" to refer to an attempt to prove something, and all other verbal disputes will be called disagreements.

Disagreements can take many forms, and any attempt to classify them will invariably be somewhat arbitrary and probably incomplete. The following classification is intended to be a rough guide:

- A *factual disagreement* is a disagreement about some matter of fact.

Joe: "The stock market, as measured by the Dow Jones average, reached its highest point in 1999."

Moe: "No, it didn't peak until sometime in 2000."

- In an *interpretive disagreement*, people can agree about the facts but disagree as to their significance.

  Joe: "Current stock market problems indicate that a major recession is just around the corner."

  Moe: "No, it's just a minor slow-down in the economy, nothing major."

- In an *evaluative disagreement*, people are offering different appraisals of what something or someone is worth.

  Joe: "Verdi's operas are much better than Wagner's."

  Moe: "No, Wagner's are the best."

- In a *verbal disagreement*, people appear to have a substantive dispute, but they are really using words in different ways.

  Joe: "If a tree falls in the woods and no one is there to hear it, then it doesn't make a sound."

  Moe: "Of course it makes a sound, and we could measure the sound wave by an unmanned instrument."

  Note: this dispute is about the word "sound". Joe is in effect defining "sound" in terms of auditory sensations, whereas Moe thinks in terms of a physicist's definition.

In each of the following examples, identify the disagreement as factual, interpretive, evaluative, or verbal.

1. Mary: I just finished reading a great book, *The Old Man and the Sea*, and I would have to conclude that William Faulkner is a superb writer.

   Maggie: I agree that Faulkner is a great writer, but he didn't write *The Old Man and the Sea*. Hemingway wrote it.

2. Cynthia: A woman has the fundamental right to choose what happens to or within her body.

Daniel: That isn't true. The state can restrict abortion in the last trimester.

3. Roger: Well, if you ask me, I would say that the best way to get from here to Lake Bruin is to take Interstate 49 from Opelousas, then take the Pineville exit and follow Highway 28 to Ferriday. From there you will go north and follow the road signs.

   Judy: No, the best way to get there is to take U.S. 190 east until you get to the Highway 1 turn-off, then follow it north. It's a much more picturesque drive and doesn't take that much longer.

4. Alice: I would like some French fries with my hamburger.

   Harold: Why do you call them French fries? The potatoes were grown in Idaho and cooked in Louisiana. There's nothing French about them.

5. Boudreaux: Since there are many re-possessed houses on the market and interest rates are low, now is the perfect time for Michele to buy a house.

   Thibodeaux: I disagree. Lending agencies are going to be much stricter in their financial checks on her sources of income, property appraisals are going to be much lower, and property values may not show much appreciation for a number of years.

Recognizing Arguments

You will recall from the previous section that an argument consists of two parts: premise and conclusion. As we shall see, there are certain words or phrases that are *premise indicators*, and other words that are *conclusion indicators*. Also, the premises usually come before the conclusion, but not always. And some arguments are very tricky because they don't include any indicator words. Consider the second amendment to the U. S. Constitution: "A well regulated Militia, being necessary to the security of a free State, the right of the people to keep and bear Arms, shall not be infringed." The sentence structure and capitalization are somewhat quaint. If we give the amendment a more contemporary wording, it will sound more like an argument: "*Because* a well-regulated militia is necessary to the security of a free state, *therefore* the right of the people to keep and bear arms shall not be infringed."

Another problem is that arguments in ordinary life are not nice and neat, as they are in logic texts. They are thrown into discussions, essays, speeches, and so forth, often without any clear demarcation as to where they begin and where they end. But we will start out simple. First, you will need to be able to identify premise and conclusion in statements that are clearly identified as arguments.

Premise Indicators	**Conclusion Indicators**
Since	Therefore
Because	Thus
For	Hence
As	So
For the reason that	Consequently
As indicated by	Which explains
May be deduced from	Allows us to infer
In view of the fact that	Obviously...
Inasmuch as	Which means that
Insofar as	So it must be that
To the extent that	As a result
On the basis of	It follows that
As shown by	We may conclude that
As implied by	Which proves that
Follows from the fact that	You can see that
On account of	This shows why
Assuming that	So it has to be the case that
	I must conclude that
	Which implies that
	Points to the fact that

<u>Exercises</u>: Identify the premises and conclusions in the following passages, each of which contains just one argument.

1. You need to vote because America doesn't work if you don't vote.

2. If you want to be an engineer, you have to take calculus.

3. Forbear to judge, for we are sinners all.

4. I am convinced that chimpanzees are conscious. Why? Well, I am conscious, chimps are our evolutionary cousins, and I know evolutionary innovations don't just spring into existence full blown.

5. The proof that the world is round rather than flat is obvious to anyone watching a ship sail away from shore. The ship will disappear from sight gradually as it goes over the horizon. If the world were flat, the ship would disappear all at once when it reached the edge of the world.

6. Jo: Anyone who does not believe in a higher power is an atheist.

   Mo: Well, God must be an atheist, because there's no higher power for him to believe in.

Arguments and Explanations

The previous set of exercises should have been fairly easy, because the arguments were isolated and identified. Now we need to add a bit of complexity by looking at arguments as they really occur in ordinary discourse, not in logic texts. The way people actually argue is not so neat. They will repeat themselves. They will throw in irrelevant material. And they don't always use the argument-indicator words. Or they may use some of the argument-indicator words when what they are really constructing is an explanation, not an argument. So our next step is to learn how to determine the presence of an argument. What we have to do is "look behind" the words and attempt to judge the author's intention. If the intention to convince us of the truth of something, then the passage is probably an argument. If, however, the author assumes that the truth of the matter is not at issue, and his (her) attempt is to answer the question how or why this came to be, then the passage is probably an explanation.

Suppose, for example, my neighbor Lauren and I engage in the following dialog:

> She: Why did you put the casserole in the oven?
>
> I: I just wanted to heat it up. I need a little snack to tide me over until supper time.
>
> She: Well, dummy, why didn't you turn the oven on?
>
> I: Another senior moment. I must have been distracted when I stopped to answer the phone.
>
> In this example, I am not really trying to prove anything, just to explain behavior that would otherwise appear stupid.

<u>Exercises</u>: Determine whether each of the following passages should be classified as an argument or an explanation. Explain your answer.

1. Judge not, that ye be not judged.
2. Every literature major has to take the Shakespeare course. Since Suzy took the course, she must be a literature major.
3. When someone disappears, the police focus on the first 48 hours. Why? Because they know from experience that if the missing person is not found within the first 48 hours, that person will probably never be found alive.
4. If the President dies in office, the First Lady becomes President. So if Obama died in office, Michelle would have been our new President.
5. Most of our general rules are known to have exceptions. Thus we say in general that birds fly, but penguins don't fly, nor do chickens that have had their wings clipped.
6. Doctor to patient: Get this prescription filled today and take the medicine as prescribed. Your symptoms should be gone by the time the bottle is empty. If not, call my office for a follow-up appointment.
7. All of the so-called "green" sources of energy, such as wind power and solar power, are insufficient to generate the amount of energy that America needs. So if we're

really serious about getting rid of fossil fuels, the only alternative is nuclear power.

8. If you want your marriage to last a lifetime, you should not marry for love, because love rarely lasts a lifetime.

9. There are three large international Christian organizations that all claim to be in some sense Catholic: the Roman Catholic Church, the Eastern Orthodox churches, and the churches of the Anglican Communion. All three owe their world-wide scope in large part to the colonialism of previous centuries.

10. He that hath taken wife and child hath given hostage to fortune, for they are impediments to great enterprise.

Informal Logic

Definitions

One of the annoying things about mathematicians and scientists is that they use ordinary words in very un-ordinary ways. Rather than inventing a new word, they will give an everyday word a technical meaning. Thus, the word "function" is defined by mathematicians to mean "a magnitude so related to another magnitude that to values of the latter there correspond values of the former" (whatever that means). This specialized use of everyday words can sometimes be confusing to the rest of us who are not part of the secret brotherhood (or whatever) that created the peculiar usage.

Unfortunately, logicians also play this game. In ordinary speech, the word "formal" refers to the way you would dress for tea with the Queen, whereas "informal" designates the attire more appropriate for a crawfish boil in the back yard (normally jeans and tee shirt). But logicians use these two words to refer to two different types of errors in reasoning (known as fallacies). A formal fallacy is an error that depends on the form or structure of an argument, without regard to its content. As we shall see later, formal fallacies can be represented using mathematical symbolism, and the errors can be demonstrated using techniques similar to those used to demonstrate errors in algebra or trigonometry.

By contrast, informal fallacies are dependent upon the content of the argument, rather than its form, and on the particular way in which the argument is worded. Some logicians prefer not to include informal fallacies in logic texts, because they regard them as misuses of language rather than errors in reasoning. However, I think this distinction is rather arbitrary since our reasoning is normally done within the context of language usage. Moreover, these errors are so common, and so effective at misleading people, that I continue to recommend their study. It will not be safe to confine them to the trash bins of history until people cease to be seduced by them. I think that time is far away.

As an example of an informal fallacy, consider the following "statistical analysis":

- Fact 1: Number of physicians in the US: 700,000. Accidental deaths caused by physicians per year: 120,000. Accidental deaths per physician: 0.171 (U.S. Dept. of Health & Human Services)
- Fact 2: Number of gun owners in the US: 80,000,000. Number of accidental gun deaths per year (all age groups): 1,500. Accidental deaths per gun owner: 0.0000188
- Conclusion: Statistically, doctors are approximately 9,000 times more dangerous than gun owners.
- Fact 3: Not everyone has a gun, but everyone has at least one doctor.
- Warning: Please alert your friends to this alarming threat. We must ban doctors before this gets out of hand.
- Addendum: As a Public Health measure I have withheld the statistics on lawyers for fear that the shock could cause people to seek medical aid.

This is obviously intended as a joke, but how would we go about showing the logical errors to someone who might be inclined to take it seriously? First of all, we should note that a statistical correlation is not in itself proof of a causal relationship. Moreover, it should be obvious that this particular correlation is contrived. In general, physicians treat people who are not healthy. Many of these people would die without medical assistance. If they die with medical assistance, has the physician caused their death? Not necessarily. Also, the "analysis" fails to consider the number of people whose lives are saved by physicians. If we were looking at this issue seriously, we should compare the number of people whose lives are saved by physicians to the number of people whose deaths are caused by physicians. That comparison would no doubt put medical assistance in a much more favorable light.

Of course, arguments that are no more logical than the guns-and-doctors analysis occur quite frequently, and most of them are intended to be taken seriously. As an example, consider the claim by atheists that religious belief has no basis because no one has ever been able to construct a solid proof of God's existence. This is countered by believers who claim that God must exist because the atheists have never been able to prove his non-existence. Probably the more appropriate conclusion is that our powers of proof are rather limited, and partisans on both sides of this issue could use a bit of humility.

Informal Fallacies

Since informal fallacies are by definition informal, any attempt to classify them is at best approximate. Many of these fallacies could just as easily be classified in one way as another. Accordingly, the definitions that follow should not be considered hard and fast rules, but guidelines to help you identify errors in reasoning. The important thing is not to memorize the names but to understand the errors.

Fallacies of Ambiguity

In each of these arguments, there is a change in meaning from premise to conclusion.

1. Equivocation: the fallacy is due to a change of meaning in a word or phrase. Example: "I don't think that hospital will accept our daughter. The sign on the front says Christian Children's Hospital, but our family is Jewish." In this example, the sign is probably intended to indicate that the hospital treats children and is owned by a Christian organization, but the parent who makes the statement takes the sign to mean that only Christian children are treated at the hospital.

2. Amphiboly: the fallacy is due to an ambiguous grammatical structure. Example: "When my son Michael was 6 years old, he loved to wrestle with me before his nightly bedtime story. Once, when I came home sick and went to bed early, my wife intercepted Michael on his way to my bedroom and explained that I wasn't feeling well and wouldn't want to wrestle. 'Oh, yes, he does,' he replied; 'because I asked Daddy if he wanted to wrestle, and he

said it was the last thing he wanted to do today.'" In this example, the father is using a figure of speech to indicate that he does not want to wrestle at all, but his son thinks he means that he wants wrestling to be his final activity of the day.

3. Composition: the fallacy is caused by applying the characteristics of the individual to the group or the part to the whole. Example: "Sixty men can do a piece of work sixty times as quickly as one man. One man can dig a post-hole in sixty seconds. Therefore sixty men can dig a post-hole in one second." Mathematics cannot be applied in this simplistic fashion to a group activity. There is no practical way for sixty people to cooperate in digging a post-hole.

4. Division: the fallacy is caused by applying the characteristics of the group to the individual or the whole to the part. Example: "If the New York Yankees won the World Series, they must have the best short-stop in baseball." If the Yankees won the World Series, we might reasonably conclude that they are the best team in baseball, but we cannot conclude that each player is the best. A team is more than the sum of the skills of the individual players.

Exercises: Identify the fallacy of ambiguity committed in each of the following passages. Explain your answer.

1. Traffic accidents are increasing. Collisions between Model T Fords are traffic accidents. Therefore collisions between Model T Fords are increasing.

2. Improbable events happen almost every day. But what happens almost every day is a very probable event. Therefore improbable events are very probable events.

3. Cars do not seem to hold their value as well as pickup trucks do, so my neighbor's pickup will hold its value better than my car.

4. Lewis Carroll, in *Through the Looking Glass*: "You couldn't have it if you did want it," the Queen said. "The rule is jam tomorrow and jam yesterday, but never jam

today." "It must sometimes come to jam today," Alice objected. "No it can't," said the Queen. "It's jam every other day: today isn't any other day, you know."

5. My new wide-screen TV won't cost much, since the monthly payments are only $89.95.

6. How could your battery be bad? You just bought that car a month ago. If the car is new, the battery must be new also.

7. My kids love surfing the Web, and they keep track of their passwords by writing them on sticky notes. One day I noticed that John's password was BatmanSupermanRobin-Joker. When I asked him why it was so long, he said, "They say it has to have at least four characters."

Fallacies of Irrelevant Premises

In each argument, the premises are, strictly speaking, irrelevant to the conclusion.

1. Attacking the person (*ad hominem*): claiming that a statement should be dismissed because of who made it. Example: Someone criticized a female sports caster by saying, "What do you expect a *woman* to know about the Super Bowl!" If a statement is true, it doesn't matter who said it – it is still true. The only exceptions are self-referential statements such as "My name is Sanford Wood."

2. Two wrongs make a right (*tu quoque* or "you're another"): attempting to justify your own faults by claiming that someone else is just as bad. Example: "Sam Porkbarrel, candidate for the 5th District Congressional seat, responded to charges that he faces criminal indictment by charging that his opponent Mary X. Torter was not being entirely honest about her own arrest record." It is almost always possible to find a worse example of something or someone, but such examples are almost always irrelevant. Someone else's vices, no matter how terrible, cannot turn my vices into virtues.

3. Appeal to force (or fear): using an implied threat as a form of persuasion. Example: "I know how much you like your job here, and how difficult it would be for a man of your age to find another job, so I am sure you will want to think carefully about the information you are going to give to the external auditor." Psychological considerations, no matter how persuasive, are logically irrelevant to the issue at hand.

4. Appeal to pity: substituting emotion for reason so that people will agree with you because they feel sorry for you. Example: "Prof. Meany, I think you should consider raising my grade. I really need to keep my average up so I can get in the nursing program. My husband left me and I have 3 kids who depend on me for their sole support." As a principle of distributive justice, need is a relevant criterion in deciding who should receive food stamps, public housing, and other social welfare services. However, assignment of grades is not part of the social welfare system, so it doesn't matter how badly a student needs a good grade.

5. Appeal to popular sentiment (*ad populum*): saying that something is right because it is popular. There are two forms to this fallacy: snob appeal and the bandwagon argument. Example: "*Gone with the Wind* must be one of the best books ever written, judging from its continued popularity." Democracy is sometimes an appropriate way to decide questions of public policy, but it is rarely an appropriate way to decide questions of truth or value.

6. Appeal to tradition (this could be considered a special form of the *ad populum* fallacy): saying that something must be right because "that's the way we've always done it". Example: "Ever since critical thinking became part of the college curriculum, students have had to work exercises on fallacies. So you can be sure you will have to do it, too." An appeal to tradition is an argument against progress. It is apparently based on the naïve belief that there was once a Golden Age when people were wise and wonderful, unlike the current generation of fools. I'm old enough to know that there have always been fools among us, in roughly the same proportion as their current distribution.

7. Misuse of authority: appealing to someone as an authority outside his/her area of expertise. Example: "Jello pudding must be good for kids. Just ask Bill Cosby!" Theoretically, any question of fact is open to verification by observation and experiment. As a practical matter, however, we don't have time to test everything, so we rely upon recognized authorities to speed up the process of learning. But the scope of expertise is always limited. When my students come to me with emotional or medical problems, I have to remind them that I'm not that kind of doctor.

8. Missing the point (*ignorantio elenchi*): drawing the wrong conclusion from the evidence. Example: "Our tests have shown that the drug in question has no medicinal value whatever, certainly none for the ailments it is claimed to cure. We conclude, therefore, that it cannot be sold successfully and will be a commercial failure." Just because a product is not useful doesn't mean it won't sell. There are plenty of people who have more money than good sense.

9. Utter irrelevance (*non sequitur*): the premises have nothing to do with the conclusion. Former President Reagan was known for this. On one occasion he observed (correctly) that "Veterans have always had a strong voice in our government." Then he added the non sequitur: "It's time to give them the recognition they so rightly deserve." Arguments like this are so dumb that no one should be deceived by them. However, I have learned over the years never to underestimate the depths of human stupidity. "Two things are infinite, the universe and human stupidity, and I'm not sure about the universe."

Exercises: Identify the fallacy of irrelevant premises committed in each of the following passages. Explain your answer.

1. An argument to show that socialism is a good economic policy: Look at the facts. At one time all utilities were privately owned; now more and more of them are owned by the government. The social-security laws embody many of the principles that socialists have always maintained. We are

well on our way to socialism, and its complete triumph is inevitable!

2. Old Man Brown claims he saw a flying saucer land on his farm. But Old Man Brown never got beyond the fourth grade in school and can hardly read or write. You can't believe what someone like that says.

3. Any person with half a brain can see that global warming is the greatest threat to mankind in the history of the world.

4. Candidate Bill Clinton, on the campaign trail: "President Bush has accused my economic recovery plan as being one of 'smoke and mirrors.' Well, if anyone should know, he would be the one."

5. Candidate George Bush: "How can you trust the governor of a small, failed state?"

6. The psychiatrists said that in their opinion Boudreaux doesn't remember the attacks after indulging in alcohol, and therefore he is not responsible for his crimes.

7. Employee to supervisor: "I definitely deserve a bonus and a raise. I'm behind on my house payments and my wife just lost her job. Not only that, but Junior needs braces and tutoring. Things are tough all over. Also, you might remember that I could take what I know to our competitors and they could bury you tomorrow."

8. What do you mean, "Why do we play the national anthem before the ballgame?" What a stupid question! We've always done it that way.

9. Teen-age son to parents: "It's true that Suzy and I lied to you about where we were going last night, but, hey, we lied to her parents, too."

Fallacies of Inadequate Premises

In these arguments, the premises provide some evidence for the conclusion, but not enough.

1. Accident: taking something that is true as a general rule and misapplying it to an exceptional case. Example: "Officer O'Malley should be arrested, because he was exceeding the speed limit in pursuit of the bank robber." The principles of logic and mathematics may be the same always and everywhere, but our laws and policies are acknowledged to have a more limited scope.

2. Converse accident (also called hasty conclusion): generalizing on the basis of insufficient evidence or samples that are not representative of the group. Example: In the Presidential election of 1948, Dewey was projected an easy winner on the basis of telephone polls. But at the time, most of the people with telephones were Republicans, so the poll was not representative of the voters.

3. Appeal to ignorance: claiming that something is true because it hasn't been proven false. Example: "There is absolutely no way that Ronald Reagan could have been involved in the Iran-Contra scandal. If so, someone would surely have been able to prove it by now." In the American legal system, a person is presumed innocent until proven guilty, but that doesn't mean that he is actually innocent. Similarly, lack of evidence that something is false doesn't mean that it is true.

4. Begging the question (*petitio principii* or circular argument): assuming as true what you are trying to prove. Example: "We know the Bible is the Word of God because it says so, and the Word of God would not lie." Circular arguments often seem to go somewhere because they use different language to say the same thing. The fallacy common to all of them is that they assume the truth of what they are trying to prove.

5. False cause: claiming that A is the cause of B just because A preceded B (*post hoc ergo propter hoc*) or, more generally, misidentifying the cause of something (*non causa pro causa*) Example: A country bumpkin was taking his first train ride. A vendor was going through the train selling fruit. The guy bought a banana, which he had never seen before. Just as he ate it, the train went into a pitch-black tunnel. The guy

concluded that eating bananas causes temporary blindness. This example illustrates what is common to many false cause arguments: people will see a causal connection in what is really coincidence. This is especially true for people who look for "signs".

6. Questionable analogy ("comparing apples and oranges"): justifying your conclusion on the basis of a comparison between two things that are more different than they are alike. Example: "When you are boiling mad, don't hold it in. If the tea kettle couldn't let off steam it would explode. So if you don't let your anger out when you're mad, you'll blow up too." As we shall see later, not all analogical arguments are fallacies. We will return to the discussion of argument by analogy when we discuss inductive inference.

Exercises: Identify the fallacy of inadequate premises committed in each of the following passages. Explain your answer.

1. Last night I got so angry at my brother that I wished he was dead. And now he's in the hospital. God, if only I hadn't thought that. It's all my fault.

2. Your Honor, I protest! This is a free country, founded on the principle that the citizens cannot be deprived of their rights and liberties. Therefore you cannot have me locked up, because that would deprive me of my rights and liberties.

3. The economy started to improve after President Trump was elected. I am certain that he must be the reason why we have turned the corner on this depressed economy.

4. Our academic committee really liked the woman we interviewed today for a teaching position. The file she submitted to support her application didn't have much in it, but we think she must be a fine teacher since there's nothing in the records to say that she's not.

5. In a Miss Universe contest several years ago, one of the contestants was asked, "If you could live forever, would you? And why?" She answered, "I would not live forever,

because we should not live forever, because if we were supposed to live forever, then we would live forever, but we cannot live forever, which is why I would not live forever."

6. If we have to teach the Biblical account of creation along with the theory of evolution, then we ought to teach the stork theory of childbirth along with the biological theory.

7. Capital punishment for repeatedly convicted drug dealers is absolutely justifiable because people who are found guilty of selling drugs again and again should be given the death penalty.

8. Treating addiction to heroin with methadone is like treating addiction to scotch with bourbon.

9. It seems to me that some kinds of extrasensory perception have to exist. In all these years no one has ever proven conclusively that the reports of its occurrence are unfounded.

Fallacies of Unfair Premises

These so-called arguments are little more than rhetorical devices designed to persuade (or trick) the listener.

1. Poisoning the well: attempting to discredit something before giving it a fair hearing. Example: "OK, let's talk about strip-mining. We all know that it's disgusting, degrading, and produces nothing but eye-sores. But let's discuss the pros and cons." Such a preface to a discussion is obviously unfair, but it is often effective because it puts your opponent on the defensive.

2. Straw man: unfairly describing your opponent (or your opponent's position) in a way that makes it vulnerable to criticism. Example: People who wish to discredit television evangelism will use Jimmy Swaggert as a typical example. This fallacy is similar to the previous one, as both attempt to prevent a fair hearing of the issue at hand.

3. Red herring: using humor or some other diversion to distract attention from the issue at hand. Example: When asked whether he thought Nancy Reagan had too much influence

on her husband's policies, Senator Bob Dole chuckled and replied that "Well, sometimes we're all a bit *too* married." This is not so much an argument as an attempt to avoid one.

4. Complex question: asking a question that assumes an unasked question has been answered in a way that is favorable to your position. Example: "Have you stopped beating your wife?" This assumes an affirmative answer to the unasked question, "Did you ever beat your wife?" This could be a diversionary technique, an attempt to avoid the issue at hand, but more often it is used to try to get the best of an opponent in a debate.

5. False dilemma: trying to convince your audience that they must choose between two extreme views when there are actually intermediate positions. Example: "America – love it or leave it!" The implied message is that if you don't like America the way it is, you should leave, because discontent shows lack of patriotism. But you could love your country dearly and still want to change it.

6. Slippery slope (also called the domino theory): claiming that if position A is adopted, then B, C, etc. will invariably follow, and each will be worse than the preceding one. Example of an argument used during the Vietnam war: "If we don't stop the Communists in Vietnam, they'll take over all of southeast Asia and we'll wind up fighting them in Australia." Well, maybe, but maybe not. The argument assumes a worst-case scenario.

7. Accent (or stress): distorting someone else's position by misplaced emphasis. Example: "It has been claimed that to really live you must freely give. Since you are not freely giving, we should have your funeral because you are not really living and if you are not living you must be dead." This could also be regarded as a fallacy of ambiguity, because the meaning shifts from living a meaningful life to simply staying alive.

8. Persuasive definition: defining yourself into victory in a debate, either by using a highly favorable definition of your

viewpoint or a highly unfavorable definition of the other person's point of view. Example: An argument against socialism began with this definition: "Socialism is theft by the poor and the lazy from the rich and industrious members of society." This is similar to poisoning the well in that it attempts to avoid a fair hearing of the issue, but the mechanism used is different.

Exercises: Identify the fallacy of unfair premises committed in each of the following passages. Explain your answer.

1. I really don't think the Federal Reserve should be lowering our interest rates by a quarter of a percent right now. People will expect them to drop it another quarter of a percent next month, and another quarter the month after that. Before you know it, the interest rate will be less than two percent.

2. If we are going to buy a car we have to buy either a good one or a cheap one. We can't afford a good one and we don't want a cheap one. So we'll just have to do without a car.

3. Feminists say they want equal pay for equal work, but this is just their opening argument. Next they will insist that 50% of the President's Cabinet must be women. It won't be long before they will demand that 50% of the players on an NFL team must be women, and so on.

4. A woman was being tried for murdering her own child. A witness stated that she had heard the woman say, "I killed my baby." When the defendant took the stand, she was asked if she had made that statement. "No," she said; "what I said was, '*I*? Killed *my* baby?'"

5. President Lyndon Johnson: "I believe in the right to dissent, but I do not believe it should be exercised."

6. Behavioral differences between men and women must be caused either by heredity or by environment.

7. Opponents of tuition increases argue that since our state benefits from well-educated university graduates, the

state should pay more for higher education through taxes. They'll probably have to pay all the costs of everyone's food, housing, and health care since it could be argued that the state also benefits from a well-fed, well-housed, healthy population.

8. Prosecutor speaking to the defendant on the witness stand: "It's clear that the accident was your fault. Did you see the traffic light turn red before or after you ran it?"

9. The drug reform movement uses the slogan "harm reduction" to disguise its true intention. It really wants to legalize drugs, so that people can buy and consume marijuana as easily as they buy and consume alcohol and tobacco.

The exercises so far have been limited to examples from each of the four groups of fallacies. Now we will do some exercises in which they are all scrambled together. As before, you should identify the fallacy or fallacies committed in each of the passages. You should also explain your answers.

1. What right has President Trump to criticize Senator Elizabeth Warren's proposed economic policies? Just remember that Trump's policies have produced the largest deficit in the history of the country.

2. Why is higher education such a waste of time?

3. Overheard at the casino: "I just play the quarter slots. That way I can't lose much money."

4. In most civilized nations, freedom of reference is an accepted way of life. Doctors consult their reference books, lawyers refer to case studies, and so on. So there's no reason students should not be allowed to use their books and notes when they are taking examinations.

5. When you think about it, democracy is one of the most desirable forms of government because it fosters both freedom and equality, which are essential to any democratic form of government.

6. We only have two options in dealing with the Taliban: back down or fight!

7. Americans are fed up with the dishonesty and corruption of their politicians. This conclusion is borne out by a recent comprehensive survey of more than 100,000 college students.

8. Now I realize you were in surgery when I gave the examination. But I cannot let you take a make-up exam. If I do that, then people who are absent due to car accidents or illnesses in the family or just plain "had the flu" would demand make-up exams, too. Then, pretty soon anyone who was scared of taking one of my tests or had something better to do that day would be asking for make-up exams.

9. I started getting all these headaches right after I changed from tap water to bottled water. So, I am sure that the bottled water must be the cause of my headaches.

10. Why should we listen to what a Catholic priest says about marriage? How can a man who has never been married know anything about marriage?

11. I know how much you like this job, so I am sure you will reconsider sending that letter to the editor. You need to remember that you have a family to feed, and you're not getting any younger. If you lose this job, you'll have trouble finding another one.

12. Most people in town are against the All-Nude Night Club. Letters in the paper are four-to-one in opposition to it and the local neighborhoods are incensed about it.

13. My brother was out driving with his 6-year-old son Nelson when they passed a race track. The boy had never seen one, so he asked what it was. "It's a place where people go to race dogs," my brother replied. After a long pause Nelson said, "I bet the dogs win."

14. What's wrong with having a public prayer at the beginning of a ball game? We've always done it that way.

15. Said before Trump was elected: Mr. Trump is obviously unqualified to be President. He knows nothing about foreign policy and doesn't even understand how the federal government works. Therefore he can never be elected.

16. The problem, as I see it, is if we give the okay for this play, there won't be any effective way for us to censor future plays that have more nudity, violence, or offensive language. In effect, we will have given a license to pornography. We might as well legalize drugs, prostitution, and organized crime while we're at it.

17. You know why I think you should not go out of your way to do a good job around here? I'll tell you why. Doing a good job around here is like wetting your pants in a dark suit. You get a nice warm feeling temporarily but nobody seems to notice. Then you're left just feeling all wet.

18. Studying philosophy can be really dangerous. It makes you think critically, which in turn can make you skeptical of your religious beliefs. And once you've begun to lose faith in religion, it's a small step to atheism and immorality.

19. America is the wealthiest nation in the history of the world, so it's ridiculous to say that we have a poverty problem in this country.

20. An argument in defense of freedom of speech: To allow every person unbounded freedom of speech must always be, on the whole, advantageous to the community, for it is highly conducive to the interests of the community that each individual should enjoy unlimited freedom of expression.

21. Since everyone is going to die, the human race will someday come to an end.

22. When his mother returned from the grocery store, the little boy grabbed the box of animal crackers that he had begged for and spread them all over the kitchen table. When his mama asked him what he was doing, he replied, "The box

says you can't eat them if the seal is broken. I'm looking for the seal to see if it's broken."

23. The distinguished senator from Arkansas, Sen. Jubilation T. Cornpone, was accused by his opponent of misappropriating funds. Sen. Cornpone responded by quoting a Biblical passage: "He who is without sin, let him cast the first stone."

24. The teachings of the Roman Catholic Church are based on the authority of the Pope. Since the Pope is infallible, these teachings must be true, and the infallibility of the Pope is one of the Church's basic teachings.

25. From a 1983 National Rifle Association membership drive letter: "We all know their Master Plan: First, outlaw all handguns. Then register all rifles and shotguns. Make no mistake, these anti-gun and anti-hunting forces are working feverishly for the day when they can gather up your rifles, handguns, and shotguns and ship them off to gun-melting furnaces."

26. Sure, I've told you before that I believe everyone's opinion counts on moral matters like abortion. But not *everyone's* opinion. Take Sue – she sleeps around like sex is going out of style next week, or something. She's just a slut, and she's broken up at least one marriage I know about. Why should her opinion on the abortion issue count for anything?

27. While General Grant was winning battles in the West, President Lincoln received many complaints about Grant's drinking problems. Lincoln is said to have replied, "I wish General Grant would send a barrel of his whiskey to each of my other generals!"

28. President George W. Bush was quoted as saying, "When I have been asked who caused the riots and the killings in L.A., my answer has been direct and simple. Who is to blame for the riots? The rioters are to blame. Who is to blame for the killings? The killers are to blame."

29. You can't believe what Professor Threadbare says about the importance of higher salaries for teachers. As a teacher himself he would naturally be in favor of increasing teachers' pay.

Formal Logic

What logicians mean by the *form* of an argument is the pattern of words and phrases that make up the argument. Sometimes we can convince someone that an argument is valid by showing that it has the same form as an argument that the person already believes to be valid. Consider this argument:

Killing a human being is wrong.
Abortion is killing a human being.
Therefore abortion is wrong.

If the validity of this argument is questioned, we can show that it has the same form as an argument known to be valid:

A dog is a mammal.
Fido is a dog.
Therefore Fido is a mammal.

It is appropriate at this point to introduce the distinction between *validity* and *soundness*. A valid argument is one in which the inference from premises to conclusion is correct. We say in such a case that the conclusion follows from the premises. But an argument can be valid and not be sound. Consider another argument that has the same pattern as the Fido argument:

A cat is a fish.
Frisky is a cat.
Therefore Frisky is a fish.

This argument, though valid, is not sound. It is not sound because one of its premises is false. So, if we return to the abortion argument, some people would question its soundness because they do not believe that the second premise is true.

To summarize:

- A *statement* is true or false. Note that different statements can have the same meaning: "It's raining", "Il pleut", "Es regnet". Statements that have the same meaning are said to express the same proposition. If a statement is true, any statement expressing the same proposition is true. Conversely, if a statement is false, any statement expressing the same proposition is false.

- *Arguments* are valid or invalid. If an argument is valid, any argument having the same form will also be valid. Conversely, if an argument is invalid, any argument having the same form will also be invalid.
- A valid argument with all of its premises true is said to be *sound.* An argument can be unsound if it is invalid or if any of its premises is false. Note that a valid argu ment with true premises will always be sound because it will have a true conclusion.

We've wandered a bit from our discussion of formal logic. Before the digression on validity and soundness, I was explaining how we could convince someone of the validity of an argument by showing that the argument in question has the same form as an argument known to be valid. Conversely, we can show that an argument is invalid by showing that it has the same form as an argument known to be invalid. Consider the infamous "guilt by association" fallacy. In one of its forms, it goes like this:

> Communists favor a redistribution of wealth from the rich to the poor.
> So do Democrats.
> Therefore Democrats are Communists.

But this argument has the same form as the obviously invalid argument:

> Cats are mammals.
> So are dogs.
> Therefore dogs are cats.

In the Middle Ages, during the long, dark winters in northern Europe, when the monks weren't busy entertaining themselves in the wine cellar, they identified and classified the various formal fallacies. The one above is called the Fallacy of the Undistributed Middle Term. But I'm getting ahead of myself.

Standard Form

Before we can use the monks' classification scheme, we need to be able to "translate" arguments from the way they are stated in ordinary English (or French or German or whatever) into a form that will permit a more careful analysis. We will begin with what logicians call *categorical propositions.* A categorical proposition is one that says categorically (no ifs, ands, or buts) that something is (or is not) something else, for example:

> Fido is a dog.
> Fido is not a cat.

The form that will permit more careful analysis is known as standard form. When a categorical proposition is in standard form, it consists of four parts:

- a *quantifier*: "all", "no", or "some";
- a *subject term*: a noun or noun phrase;
- a *copula*: "are" or "are not";
- a *predicate term*: a noun or noun phrase.

Examples: All dogs are mammals.
No bachelors are married men.

There are four types of categorical propositions, conventionally designated as follows:

- Type A: universal affirmative ("All ... are").
- Type E: universal negative ("No ... are").
- Type I: particular affirmative ("Some ... are").
- Type O: particular negative ("Some ... are not").

To call attention to the form of an argument, as opposed to its content, it will be useful to symbolize the propositions. By convention, capital letters are used to designate the subject and predicate terms, with one of the type letters (a, e, i, or o – but not u) between the letters for the two terms. Using the examples above:

> "All dogs are mammals." would be symbolized as DaM.
>
> "No bachelors are married men." would be symbolized as BeM.

Exercises: Translate the following statements into standard-form categorical propositions. Symbolize the propositions using SaP format. The first three are done for you as examples.

1. Roses are fragrant.
   Translation: All roses are fragrant things. (RaF)

2. All that glitters is not gold.
   Translation: Some things that glitter are not golden things. (GoA)

   Note: This is not a universal statement, even though it begins with "all". A more accurate way to say it would be, "It's not true that everything that glitters is gold", from which we can infer that some things that glitter are not golden things. Note also that I used A for gold because I could not use G twice, and I remembered Au as the chemical abbreviation for gold.

3. Only members can use the front door.
   Translation: All people who can use the front door are members. (FaM)

   Note: The general rule for statements beginning with "only" is to replace "only" with "all" and interchange subject and predicate. What the statement says is that if you're not a member, you can't use the front door, from which we can infer that if you can use the front door, you must be a member.

4. Some advertisements are useful in bringing buyers and sellers together.

5. No people over 65 are eligible to claim the earned-income tax credit.

6. Many research scholars are very poor teachers.

7. Most of my recipes are modifications of recipes that I have gotten from other people.

8. There are no computer networks that are completely secure.

9. Harry only has a good time when he has lots of money to spend.

10. There are some possible courses of action which have been overlooked.
11. Some of the most intelligent and sensitive physicians are not board-certified.
12. Dogs with long hair stay warmer in winter than people with long hair.
13. Almost no one walks ten miles to school uphill in the snow anymore.
14. Cayenne pepper is tasty but not fragrant.
15. Many a person has lived to regret a misspent youth.
16. Not everyone worth meeting is worth having as a friend.
17. Very few things are both safe and exciting.
18. Only the good die young.
19. Good advice is not always appreciated.
20. The fish never bite after a rain.
21. No easy exercise ever gives me a headache.
22. He who takes the sword shall perish by the sword.
23. A thing of beauty is a joy forever.
24. Whatsoever a man soweth, that shall he also reap.

Immediate Inference

The term "immediate inference" refers to what can be inferred from a single proposition, without the addition of any other information. In effect, we have an argument with a single premise.

By way of introduction to this topic, let's think about what we mean when we say that two statements are opposite in meaning. Our ordinary use of the word "opposite" is not very precise. We can make our meaning more precise by a series of definitions. For the present, we will restrict our discussion to categorical propositions that are in standard form.

- Such propositions can be opposite in quality, that is, either affirmative or negative.
- They can be opposite in quantity, that is, either universal or particular.
- They can be opposite in truth-value, that is, either true or false.

If we use S to represent the subject term and P to represent the predicate term, we can show the relationships among propositions having the same subject and predicate terms by means of the following diagram. This diagram has traditionally been called the *Square of Opposition,* although some of my students dubbed it the Square of Confusion.

The propositions on opposite corners of the square (diagonally) are called *contradictories.* They are completely opposite, that is, they are opposite in quantity, quality, and truth-value. That means we can infer the falsity of one from the truth of the other, or vice versa.

Example: "All dogs are mammals" is true.

Therefore "Some dogs are not mammals" is false.

Example: "No dogs are mammals" is false.

Therefore "Some dogs are mammals" (meaning "at least some") is true.

The propositions across the square from each other at the top and the bottom are not completely opposite. They are opposite in quality but not in quantity. First, consider the two propositions at the top: SaP and SeP. They are called *contraries*. They cannot both be true, but they could both be false. So if we know that SaP is true, we can infer that SeP is false.

Example: "All dogs are mammals" is true.

Therefore "No dogs are mammals" is false.

If, however, SaP is false, the truth-value of SeP cannot be determined.

Example: "All dogs are cats" is false.

But "No dogs are cats" cannot be inferred.

Because: "All dogs are black" is false.

But "No dogs are black" cannot be inferred.

As the previous example shows, sometimes both contraries are false, sometimes not. That is why no inference can be made if all we know is that one of them is false.

The propositions at the bottom of the square are called *sub-contraries*. They cannot both be false, but they could both be true. So if we know that SiP is false, we can infer that SoP is true.

Example: "Some dogs are cats" is false.

Therefore "Some dogs are not cats" (meaning "at least some") is true.

If, however, SiP is true, the truth-value of SoP cannot be determined

Example: "Some dogs are black" is true.

But "Some dogs are not black" cannot be inferred.

Because: "Some dogs are mammals" (meaning "at least some") is true.

But "Some dogs are not mammals" cannot be inferred.

As the previous example shows, sometimes both sub-contraries are true, sometimes not. That is why no inference can be made if all we know is that one of them is true.

Inference Along the Sides of the Square

All of the opposites we have looked at so far have been on the opposite side of the square. They were all opposite in quality (and, in the case of contradictories, they were opposite in quantity also). If propositions are on the same side of the square (A and I or E and O), they are opposite in quantity but not in quality. Using what we have already learned, we could determine their truth-value by a circuitous route. For example, if SaP is true, then SeP must be false; and if SeP is false, then SiP must be true.

To avoid this circuitous route, logicians have coined a name for the propositions on the same side of the square. The propositions on the top are called super-alterns and those below are called sub-alterns. The truth-values are as follows:

- The truth of the super-altern implies the truth of the sub-altern.
- The falsity of the sub-altern implies the falsity of the super-altern.
- Everything else is undetermined.

Example:

If it's true that all dogs are mammals, then it's true that at least some are;

but if it's false that all dogs are black, we cannot infer that it's false that some are.

Example:

If it's false that some dogs are cats, it must be false that all are;

but if it's true that some dogs are black, we cannot infer that all are.

Now let's try some exercises: Each problem should be worked two ways. We will first assume that the first statement in each set is true, then try to determine whether the second statement is true or false. Then we will repeat the process, assuming that the first statement is false.

1. All persons in attendance were carrying wallets.

   Some persons in attendance were not carrying wallets.

   First, we will symbolize the propositions as PaW and PoW. A and O propositions are diagonally across the square from each other, so they would be contradictories. If PaW is true, then PoW must be false; but if PaW is false, then PoW must be true.

2. None of the cookies were made with chocolate chips.

   All of the cookies were made with chocolate chips.

   We cannot use C to represent both cookies and chocolate chips, so we will use H for chocolate chips. Symbolizing the propositions as CeH and CaH, we see that they are contraries. If it is given that CeH is true, then CaH must be false; but if CeH is false, the truth-value of CaH cannot be determined.

3. Some children are not over 5 feet tall. (CoO)

   No children are over 5 feet tall. (CeO)

   CeO is the super-altern of CoO. Using the rules for alterns, we can say that if CoO is true, the truth-value of its super-altern cannot be determined; but if CoO is false, then CeO must also be false.

   Now let's go the circuitous route to show that we can get the same result. If CoO is true, its contradictory CaO must be false; but from the falsity of CaO nothing can be inferred about the truth-value of its contrary CeO. However, if CoO is false, its contradictory CaO must be true, and its contrary CaO must be false. An alternative circuitous route is to go along the bottom of the square. If CoO is false, its sub-contrary CiO must be true, and its contradictory CeO must be false. There's more than one way to skin a cat!

<u>Exercises</u>: Symbolize the following statements in SaP format. In each numbered group, explain the relationship of the second statement to the numbered statement above it (contradictory, contrary, etc.). Assume first that the numbered statement is true, and state whether the other statement is true, false, or unknown. Repeat the process, assuming that the numbered statement is false.

1. Some rap singers are true artists.
   Some rap singers are not true artists.
2. Some accountants are not punctual.
   All accountants are punctual.
3. All salesmen met their quotas.
   Some salesmen met their quotas.
4. No sailboats are over 300 feet long.
   Some sailboats are over 300 feet long.
5. Some knowledge is a dangerous thing.
   All knowledge is a dangerous thing.
6. All wizards are magicians.
   No wizards are magicians.
7. Some people are not popular.
   Some people are popular.

Other Immediate Inference

There are some propositions that cannot be compared directly on the square of opposition because they do not have the same subject and predicate terms. Consider, for example, "No dogs are cats" (DeC) and "Some cats are dogs" (CiD). It is intuitively obvious that these statements are opposite, but we cannot show their opposition by use of the square of opposition alone. We need additional tools.

One of these additional tools is called *conversion.* A proposition can be converted by interchanging its subject and predicate terms. Thus, the converse of "No dogs are cats" would be "No cats are dogs". The converted proposition (CeD) can be compared with CiD on the square of opposition. Since they are contradictories, they would be opposite in truth-value.

Converses are not always equivalent; that is, we cannot infer the truth of one from the truth of the other. E and I propositions are equivalent when converted; A and O propositions are not. For example, "All dogs are mammals" is true, but its converse, "All mammals are dogs" is not true. We will use a two-way arrow to show that two propositions are equivalent in truth-value:

SeP ↔ PeS or SiP ↔ PiS

Exercises: Symbolize the following statements in SaP format. Convert each statement, then say whether the converse is logically equivalent.

1. No electricians are belly-dancers.
2. Pit bulls sometimes turn on their owners.
3. Almost all kittens are lovable.
4. Some cans are not recyclable.
5. Most people who love classical opera do not like rap music.
6. Only the very best singers are selected for the Metropolitan Opera.
7. People who came late to the polls did not get to vote.
8. Not everyone who is registered to vote did vote in the last election.
9. Where there's smoke there's fire.
10. No sane witnesses will incriminate themselves.

Another tool is called *obversion*. Obversion is a double negative. It involves changing the quality of a proposition and replacing the predicate with its complement. The word "complement" refers to things that are not members of the class designated by the term in question. Thus, the complement of cats is everything that's not a cat, the complement of people over six feet tall is people not over six feet tall, etc. Even though it sometimes sounds awkward, it is conventional to use the prefix "non" to designate the complement of a class: non-union members, non-strikers, non-cats, etc. We will represent the complement of a class by drawing a bar over the letter that designates the class.

Examples:

"All dogs are mammals" becomes "No dogs are non-mammals": $DaM \leftrightarrow De\overline{M}$

"No dogs are cats" becomes "All dogs are non-cats": $DeC \leftrightarrow Da\overline{C}$

Note that obverses are always equivalent.

Exercises: Symbolize the following statements in SaP format, then obvert them.

1. Most beignets are tasty treats.
2. No wart hogs are lazy beasts.
3. Only natural-born citizens can run for President.
4. Some bananas are not ripe enough to make banana pudding.
5. Every valid scientific experiment is repeatable.
6. Polluted cities are not desirable places to live.
7. Southern states usually vote Republican in the Presidential race.
8. Some automobiles are electrically powered.
9. No one who heard the guest speaker remained unmoved.
10. Not every registered Republican voted for Trump.

Another kind of double negative is the *contrapositive*. Contraposition involves replacing the subject by the complement of the predicate, and the predicate by the complement of the subject.

Example: "Every dog is a mammal" becomes "Any animal that's not a mammal is not a dog" or "All non-mammals are non-dogs".

Symbolically: $DaM \leftrightarrow \overline{M}a\overline{D}$

Contraposition produces equivalent propositions only for A and O propositions. The reason for this is apparent if we see how a contrapositive is derived. It is really the obverse of the converse of the obverse: $DaM \leftrightarrow De\overline{M} \leftrightarrow \overline{M}eD \leftrightarrow \overline{M}a\overline{D}$

If the initial proposition was E or I, the derivation process would involve an illegal conversion.

Now that we have a full set of tools at our disposal, it's time to do some more complicated exercises. In each of the following problems, we have a given proposition followed by several additional propositions which are to be compared to the original proposition either on the square of opposition or by some combination of conversion, obversion, and contraposition. As in the previous set of exercises, we are to assume first that the original proposition is true, then that it is false.

1. All socialists are pacifists. (SaP)

   - Some non-pacifists are not non-socialists. ($\bar{P}o\bar{S} \leftrightarrow SoP$)

   By taking the contrapositive of this proposition we get SoP, which is the contradictory of the original. If SaP is true, its contradictory SoP is false, and so is its contrapositive. If SaP is false, its contradictory is true and so is its contrapositive.

   - No non-pacifists are socialists. ($\bar{P}eS \leftrightarrow Se\bar{P} \leftrightarrow SaP$)

   By converting and obverting this proposition, we see that it is equivalent to the original. Thus, its truth-values will be the same as those of the original proposition.

   - Some pacifists are not non-socialists. ($Po\bar{S} \leftrightarrow PiS$ $\leftrightarrow SiP$)

   This time we obvert first, then convert. The resulting proposition, SiP, is the sub-altern of the original. If the original (SaP) is true, its sub-altern is true, and all of the propositions in the equivalence chain would also be true. If the original is false, the truth-value of the sub-altern cannot be determined.

   - All non-socialists are non-pacifists. ($\bar{S}a\bar{P} \leftrightarrow PaS$)

   Now we are stumped. We have a proposition that needs to be converted, but conversion is illegal for an A proposition. Thus there is no basis of comparison to the original proposition, and no way to determine the truth-value of this proposition. (I should point out that some textbooks describe a process called "conversion by limitation" for A propositions. What they are referring to is the converse of the sub-altern: PiS $\leftrightarrow$

SiP. But that does us no good. If SaP is true, its sub-altern SiP is also true, and so is its converse PiS. So far this looks promising, but now we see that our efforts were in vain, because from PiS we can infer nothing about the truth-value of its super-altern PaS. Similarly, if the original (SaP) is false, the truth-value of its sub-altern cannot be determined. The upshot of this discussion is that "conversion by limitation" is excess intellectual baggage that does us no good. Henceforth we will ignore it.)

2. Some merchants are not pirates (MoP).

   - Some pirates are merchants. (PiM $\leftrightarrow$ MiP)

   The converse of this proposition is the sub-contrary of the original. If the original is true, the truth-value of its sub-contrary cannot be determined. If the original is false, its sub-contrary is true and so is its converse.

   - No non-pirates are merchants. ($\overline{\text{P}}$eM $\leftrightarrow$ Me$\overline{\text{P}}$ $\leftrightarrow$ MaP)

   By first converting and then obverting this proposition we get MaP, which is the contradictory of the original. Thus, the truth-values of this proposition will be the opposite of those of the original proposition.

   - No non-merchants are pirates. ($\overline{\text{M}}$eP $\leftrightarrow$ Pe$\overline{\text{M}}$ $\leftrightarrow$ PaM)

   Love's labor's lost on this one. We have an A proposition that needs to be converted, so we're stuck. Since the original is an O proposition, it cannot be converted either. There is no basis of comparison.

<u>Exercises:</u> Each of the following sets of exercises should be worked in two ways. First, assume that the original proposition (0) is true, and determine what may be inferred about the truth or falsehood of the other propositions. Repeat the process, assuming that the proposition 0 is false.

0. All socialists are pacifists.
1. No socialists are non-pacifists.
2. No non-socialists are non-pacifists.
3. All non-pacifists are non-socialists.
4. No pacifists are non-socialists.

5. Some socialists are not pacifists.
6. All pacifists are socialists.
7. Some non-pacifists are socialists.

0. No scientists are philosophers.
1. No non-philosophers are scientists.
2. Some non-philosophers are not non-scientists.
3. All non-scientists are non-philosophers.
4. No scientists are non-philosophers.
5. No non-scientists are non-philosophers.
6. All philosophers are scientists.
7. Some non-philosophers are scientists.
8. All non-philosophers are non-scientists.
9. Some scientists are not philosophers.
10. No philosophers are non-scientists.

0. Some saints were martyrs.
1. All saints were martyrs.
2. Some non-martyrs were not non-saints.
3. No non-saints were martyrs.
4. Some non-martyrs were saints.
5. Some martyrs were not non-saints.
6. No martyrs were non-saints.
7. Some non-saints were not non-martyrs.
8. All martyrs were saints.
9. No saints were martyrs.
10. All martyrs were non-saints.
11. Some non-saints were not martyrs.
12. No non-martyrs were saints.
13. No saints were non-martyrs.
14. Some non-martyrs were non-saints.
15. No martyrs were saints.
16. Some non-saints were non-martyrs.
17. No non-martyrs were non-saints.
18. Some non-saints were martyrs.
19. All non-martyrs were saints.
20. Some saints were not non-martyrs.
21. Some martyrs were not saints.
22. No non-saints were non-martyrs.
23. Some martyrs were saints.
24. Some saints were non-martyrs.
25. All non-martyrs were non-saints.
26. All saints were non-martyrs.

27. Some saints were not martyrs.
28. All non-saints were non-martyrs.
29. Some martyrs were non-saints.
30. All non-saints were martyrs.
31. Some non-martyrs were not saints.

0. Some merchants are not pirates.
1. Some non-pirates are not non-merchants.
2. No pirates are non-merchants.
3. All merchants are non-pirates.
4. All non-pirates are non-merchants.
5. No merchants are pirates.
6. No non-merchants are non-pirates.
7. All non-pirates are merchants.
8. All non-merchants are pirates.
9. Some pirates are not non-merchants.
10. No non-pirates are non-merchants.
11. Some merchants are pirates.
12. Some pirates are not merchants.
13. All pirates are non-merchants.
14. Some merchants are not non-pirates.
15. Some non-pirates are non-merchants.
16. Some merchants are non-pirates.
17. Some non-pirates are merchants.
18. Some non-merchants are not pirates.
19. Some non-merchants are not non-pirates.
20. All non-merchants are non-pirates.
21. Some non-merchants are pirates.
22. Some pirates are non-merchants.
23. No merchants are non-pirates.
24. Some non-pirates are not merchants.
25. All merchants are pirates.
26. No pirates are merchants.
27. Some non-merchants are non-pirates.
28. All pirates are merchants.

Immediate Inference and Translation

Now let's make things a little more interesting by using propositions that require a bit of translation.

Example: "Most lawyers are rich." In standard form this becomes "Some lawyers are rich people." (LiR)

- "Some attorneys are not rich."

  Translating, we get "Some lawyers are not rich people." (LoR). This is the sub-contrary of the original. If the original is true, this proposition is undetermined; if the original is false, it is true.

- "Some wealthy people are not lawyers."

  Substituting "rich" for "wealthy", we get RoL. This proposition cannot be converted but the original can be: LiR ↔ RiL. This permits us to say that the current proposition is the sub-contrary of the converse of the original, and its truth-values are the same as those of the previous proposition.

- "No poor people are attorneys."

  This translates to "No non-rich people are lawyers":
  $\overline{R}eL \leftrightarrow Le\overline{R} \leftrightarrow LaR$

  By converting and obverting the translated proposition we get LaR, which is the super-altern of the original. If the original is true its super-altern is undetermined. If the original is false its super-altern is false and so are all the propositions in the equivalence chain.

By way of review, there are five types of categorical propositions whose truth-value we may attempt to determine:

1. Propositions that can be compared to the original proposition on the square of opposition.
2. Propositions that can be shown to be equivalent to the original proposition by a sequence of conversion and/or obversion.
3. Propositions that must be first converted, etc., then compared to the original on the square of opposition.

4. Propositions that cannot be compared to the original itself, but can be compared to an equivalent of the original.
5. Propositions for which there is no basis of comparison to the original.

By way of illustration, assume that our original proposition is SiP.

1. SeP → contradictory of the original.
2. $Po\bar{S}$ → obverse of the converse of the original.
3. $\bar{P}a\bar{S}$ → contrapositive of the super-altern (SaP) of the original.
4. $\bar{S}a\bar{P}$ → contrapositive of the super-altern (PaS) of the converse (PiS) of the original.
5. $\bar{S}oP$ → no basis of comparison

<u>Exercises</u>: In each of the following groups of propositions, symbolize the propositions so that they may be compared on the square of opposition, performing any necessary conversion, obversion, etc. In order to do this, you may need to use synonyms and change the sentence structure. If we assume that the first proposition in each group is true, what may be inferred about the truth or falsity of the other three propositions in the group? What if the first proposition is false?

1. Most athletes are not drinkers.
   Some non-drinkers are athletes.
   Some non-athletes are not drinkers.
   Some drinkers are athletes.
2. Some Asian nations are belligerents.
   All of the belligerent nations are outside of Asia.
   None of the non-belligerent nations are in Asia.
   All of the Asian nations are non-belligerents.
3. Some gasses are invisible.
   Nothing you can see is a gas.
   Any gas can be seen.
   Some things you cannot see are gasses.

4. Boards less than a foot long are useless.
   Most useful boards are at least a foot long.
   All boards that are at least a foot long are useful.
   Some useful boards are less than a foot long.

5. Everyone who was neither a member nor a guest of a member was excluded.

   Some of those included were neither members nor guests of members.

   If you're either a member or a member's guest, then you can be sure you're included.

   Some members and their guests were excluded.

Extended Inference

So far our discussion of formal logic has been restricted to arguments with a single premise. Now we need to move along to arguments containing more than one premise. As is customary, we will start simply and add complexity bit by bit. We will begin with a special type of two-premise argument called the syllogism.

Syllogisms

A *categorical syllogism* is an argument consisting of three categorical propositions: two premises and a conclusion. An example is: Beagles are mammals, because beagles are dogs and dogs are mammals.

First, some *definitions*:

- A syllogism must have 3 terms, each of which occurs twice. The <u>terms</u> of a syllogism are:
    - Major: the predicate of the conclusion.
    - Minor: the subject of the conclusion.
    - Middle: the term that does not occur in the conclusion.
- A categorical syllogism is said to be in *standard form* if:
    - it consists of three categorical propositions, two of which are premises;
    - each of its propositions is in standard form;
    - each term appears twice;
    - the first premise contains the major term;
    - the second premise contains the minor term;
    - the conclusion occurs last.

My example syllogism is obviously not in standard form. The individual statements are not in standard form, and they are in the wrong order. First, let's put the individual propositions in standard form:

All beagles are mammals.
All beagles are dogs.
All dogs are mammals.

Now let's put them in the right order. "All beagles are mammals" is the conclusion, so it must go last. Also, we use it to determine the major and minor terms. Since "mammals" is the predicate of the conclusion, it is the major term, and the premise containing "mammals" must go first. "Beagles" is the subject of the conclusion, so it is the minor term, and the premise containing it must go second. So now we are ready to re-state our argument in standard form:

All dogs are mammals.
All beagles are dogs.
Therefore all beagles are mammals.

This argument appears to be valid, but appearances can be deceiving. There are other arguments that appear to be valid but aren't. Many people are tricked by the type of argument that I mentioned at the beginning of the discussion of formal logic:

Communists favor a redistribution of wealth from the rich to the poor.
So do Democrats.
Therefore Democrats are Communists.

What we obviously need are some rules by which we can evaluate categorical syllogisms. However, the rules require the introduction of a new concept called *distribution.*

A term is said to be distributed if it refers to all the members of a class.

- In an A proposition, only the subject is distributed.
- In an E proposition, both subject and predicate are distributed.
- In an I proposition, neither subject nor predicate is distributed.
- In an O proposition, only the predicate is distributed.

Now for the *rules*. Note that for each rule there is a corresponding *fallacy* for an argument that violates the rule.

1. At least one of the premises must be affirmative. (Fallacy: Two Negative Premises or Exclusive Premises)

2. If a premise is negative, the conclusion must be negative. (Fallacy: Drawing an Affirmative Conclusion from a Negative Premise)

3. The middle term must be distributed at least once. (Fallacy: Undistributed Middle Term)

4. If a term is distributed in the conclusion, it must be also distributed in a premise. (Fallacy: Illicit Process)

   Note: Sometimes the "offending" term will be identified and two names will be used: Illicit Minor or Illicit Major.

5. If the conclusion is particular, one of the premises must be particular. (Fallacy: Drawing a Particular Conclusion from Universal Premises)

Now let's work an example:

> All chocolate éclairs are fattening foods, because all chocolate éclairs are rich desserts, and some fattening foods are not rich desserts.

First, we symbolize the conclusion: CaF. Since F is the major term ("fattening foods"), the major premise is "Some fattening foods are not rich desserts", symbolized as FoR. The remaining premise, "All chocolate éclairs are rich desserts", must be the minor premise. We can symbolize it as CaR. Putting the argument in standard form, we have:

FoR
<u>CaR</u>
CaF

This is clearly an example of the fallacy of Drawing an Affirmative Conclusion from a Negative Premise.

Let's look at another example:

> All inventors are people who see new patterns in familiar things, so all inventors are eccentrics, since all eccentrics are people who see new patterns in familiar things.

In standard form:

EaN
IaN
IaE

Since this argument contains three universal affirmative propositions, we can eliminate the possibility of fallacies involving negative premises (rules 1 and 2), as well as the fallacy concerning a particular conclusion (rule 5). If there is a fallacy, it would have to be a fallacy of distribution. In an A proposition, only the subject term is distributed. We will indicate distribution by drawing a circle around the distributed term:

E a N
I a N
I a E

Clearly, the middle term, N, is not distributed at all, so this argument violates rule 3.

One more example:

> Some snakes are not dangerous animals, but all snakes are reptiles, therefore some reptiles are not dangerous animals.

In symbolic form:

SoD
SaR
RoD

We have a negative premise and a negative conclusion, so that's OK. We have a particular premise and a particular conclusion, so that's OK. Now we need to check for distribution:

S o D
S a R
R o D

The middle term, S, is distributed once. The major term, D, is distributed in both premise and conclusion. So there is no violation. This argument is valid.

Exercises: For each of the following arguments, symbolize the constituent statements, rearrange the statements (if necessary) so that the syllogism will be in standard form, and identify any fallacy committed.

1. Some foods that contain iron are toxic substances, for all fish containing mercury are foods that contain iron, and all fish containing mercury are toxic substances.
2. All opponents of basic economic and political changes are outspoken critics of the liberal leaders of Congress, and all right-wing extremists are opponents of basic economic and political changes. It follows that all outspoken critics of the liberal leaders of Congress are right-wing extremists.
3. No people who write dirty books are people you would want as neighbors, but some journalists are not people who write dirty books, so some journalists are people you would want as neighbors.
4. All Democrats are socialists, so all Democrats are opponents of the Republican Party, because all socialists are opponents of the Republican Party.
5. No fattening foods are nourishing foods, because all foods that are bad for you are fattening foods, and no foods that are bad for you are nourishing foods.
6. No fattening foods are nourishing foods, because no fattening foods are natural grain products, and all natural grain products are nourishing foods.
7. All people who live in London are people who drink tea, and all people who drink tea are people who like crumpets.

We may conclude, then, that all people who live in London are people who like crumpets.

Venn Diagrams

What we have seen by these examples and exercises is that the rules alone provide a rather time-consuming process for valid arguments. If an argument is invalid, we can often identify the fallacy right away. But if the argument is valid, then we have to use a process of elimination: it doesn't violate rule 1, and it doesn't violate rule 2, and so forth. We need a more efficient method of testing validity.

First, let's consider *an alternative symbolism.* We can think of the subjects and predicates of categorical propositions as referring to classes of things. But just because a class has a name (and perhaps even a definition), we cannot infer that the class has any members. Perhaps the name refers to things that used to exist but have become extinct (like *dinosaur*), or perhaps it refers to fictitious things (like *unicorn*). In any event, it makes sense to ask whether a class is empty.

An alternative way to symbolize the A, E, I, and O propositions is to say that they are equal or unequal to zero (that is, they refer to classes that are empty or not empty).

- Thus, SaP is first obverted to $\text{Se}\bar{\text{P}}$, then restated as $\text{S}\bar{\text{P}} = 0$. For example, to say that all dogs are mammals (DaM) is equivalent to saying that there are no dogs that are not mammals ($\text{De}\bar{\text{M}}$).
- SeP is re-symbolized as SP=0. For example, to say that no dogs are fish (DeF) is equivalent to saying that the class that contains dogs that are fish is empty.
- SiP says that there is at least one thing that is both S and P, that is, the class of those things that are both S and P is not empty: $\text{SP} \neq 0$.
- SoP is first obverted to $\text{Si}\bar{\text{P}}$, then symbolized as $\text{S}\bar{\text{P}} \neq 0$.

Note that universal propositions are always equal to zero, particular propositions unequal. Note also that this symbolism makes it obvious that A and O propositions are contradictories, and that E and I propositions are contradictories.

A pictorial representation of these relationships is provided by *Venn diagrams*. A Venn diagram consists of two overlapping circles: If we think of these circles as representing classes, then we can label them as S and P to represent the subject and predicate terms of a proposition.

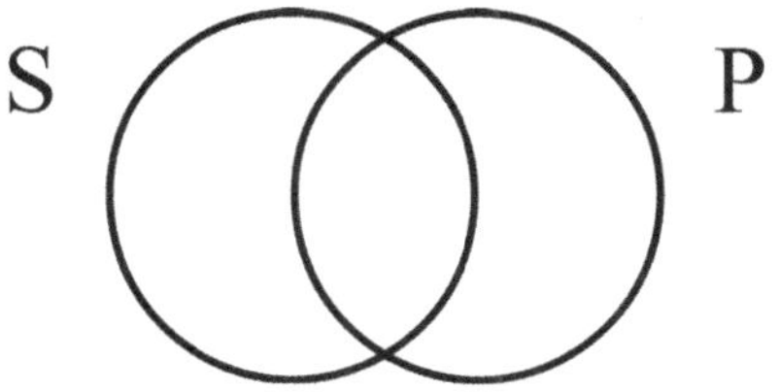

We can think of the three areas of the diagram as follows:

- The left-most area, $S\bar{P}$, the area of S that is completely outside of P.
- The middle area, SP, the area common to S and P. This area represents the intersection of the two classes.
- The right-most area, $\bar{S}P$, the area of P that is completely outside of S.

By convention, an area of the diagram is shaded to indicate that it is empty. An X is placed in an area to show that it is not empty. We can use Venn diagrams to represent the four types of categorical proposition as follows:

- SaP or $S\bar{P} = 0 \rightarrow$ S outside of P is empty.

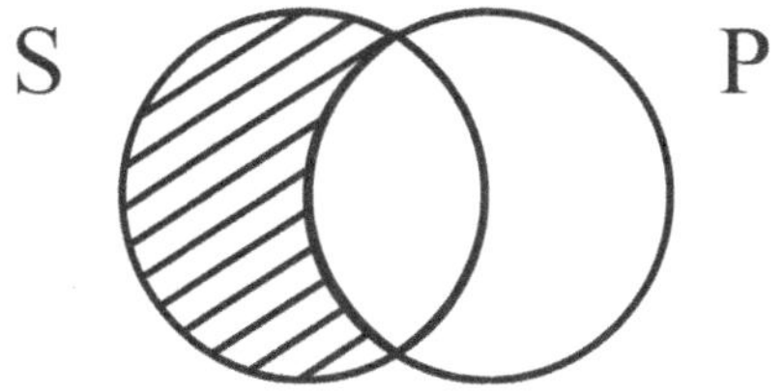

- SeP or SP = 0 → The intersection of S and P is empty.

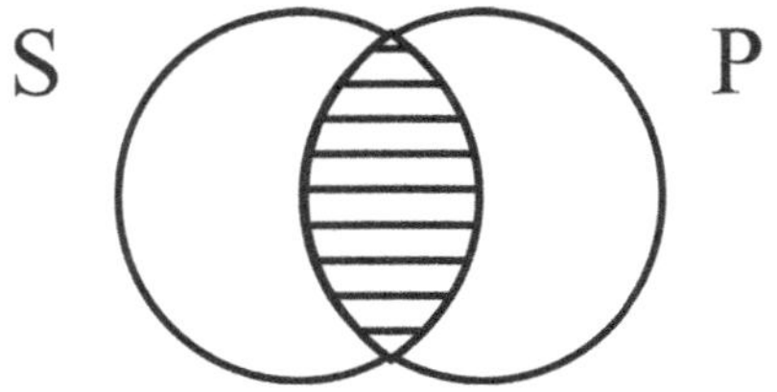

- SiP or SP ≠ 0 → The intersection of S and P is not empty.

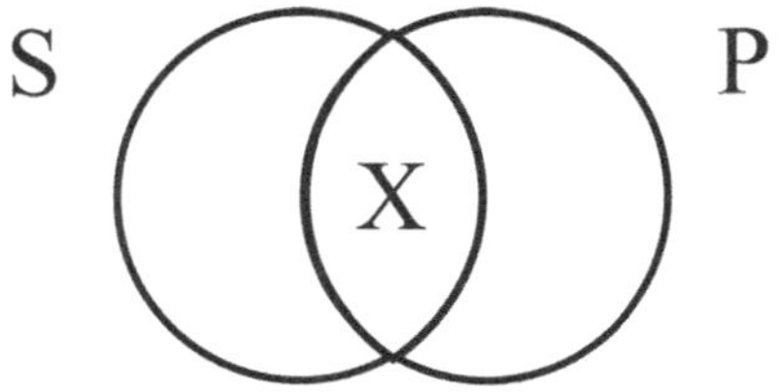

- SoP or $S\overline{P} \neq 0$ →S outside of P is not empty.

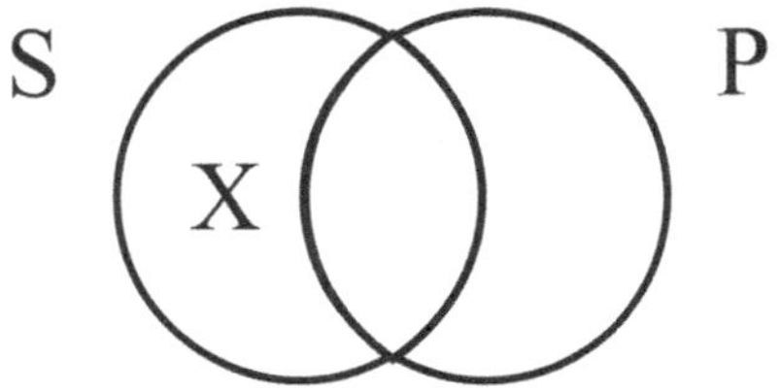

We can use a Venn diagram to test the validity of a syllogism by extending the diagram to include three interlocking circles. Note the eight areas of the diagram:

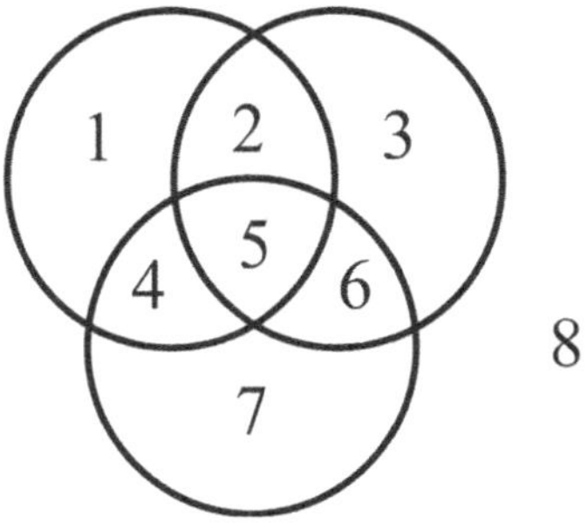

Since a syllogism in standard form must include exactly three terms, each circle can be used to represent a term. Consider the argument:

All dogs are mammals.	DaM
All beagles are dogs.	<u>BaD</u>
Therefore all beagles are mammals.	BaM

This argument can be symbolized as follows:

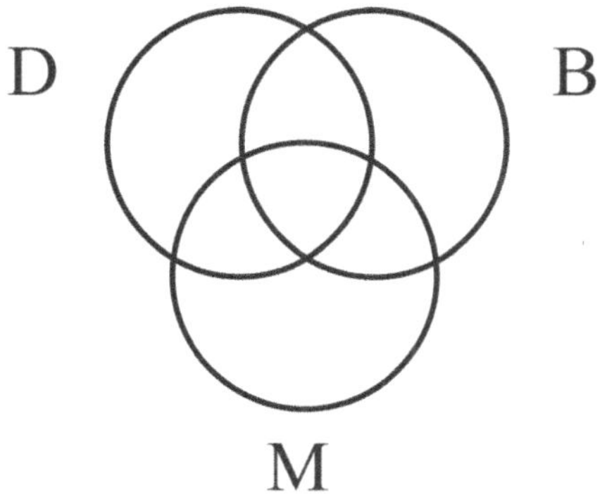

The idea behind a Venn diagram test is that a valid argument implicitly contains its conclusion in the premises. So if we diagram the premises, the conclusion should be diagrammed in the process. To diagram the major premise, DaM, we shade out the area of D outside of M. To diagram the minor premise, BaD, we shade out the area of B outside of D. If the argument is valid, then the conclusion should implicitly be diagrammed, that is, the area of B outside of M should be shaded out, and it is.

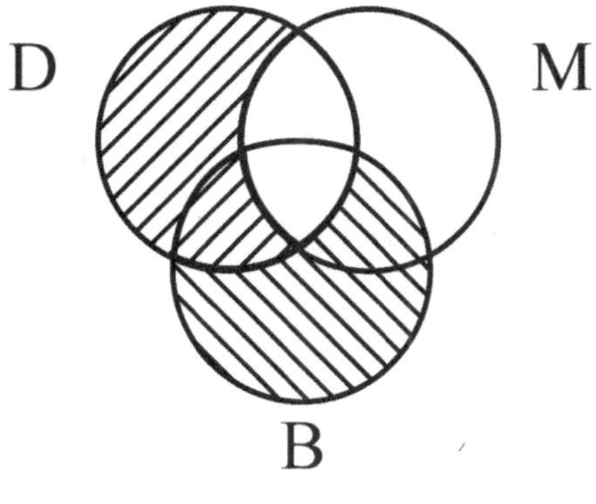

Note that we do not explicitly diagram the conclusion. If we did, every argument would appear to be valid. Now let's go back and see how the Venn diagram test works with arguments that we previously identified as invalid. This is the argument that was identified as an example of the fallacy of Drawing an Affirmative Conclusion from a Negative Premise.

FoR
CaR
CaF

In diagramming the major premise, it is not clear whether the X should be put in area 1 or area 4. Because of this ambiguity, we introduce a rule of practice: When one premise is particular and one universal, always diagram the universal premise first. This will not always remove the ambiguity, but it's worth a try. When the diagram is complete, we see that the argument is invalid. The conclusion, CaF, says in effect that if there is any C, it is in F, so all the areas of C outside of F should be shaded. However, area 6, which is inside C but outside of F, is not shaded. This shows that the argument is invalid, and we will mark the area with an arrow. Note that the diagram doesn't show what the fallacy is; we have to go to the rules for that.

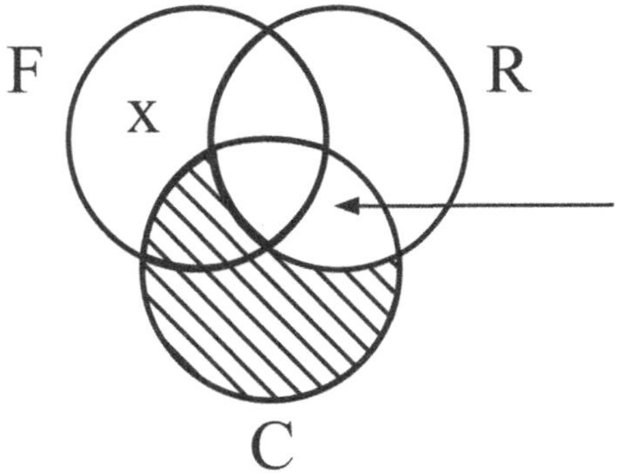

One more example: EaN
IaN
IaE

Note that in diagramming the minor premise, we "cross-hatch" area 4. This is an indication of invalidity, but not a proof. The proof of invalidity is that area 6 should be shaded but isn't.

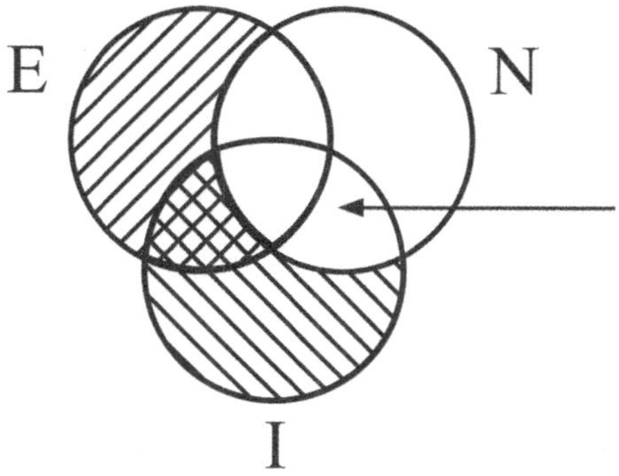

Making Arguments into Syllogisms

Many arguments that do not have syllogistic form can be made into syllogisms by one of the following techniques.

1. If an argument contains more than three terms, we may be able to reduce it to three terms by conversion, obversion, and contraposition.

2. Arguments in ordinary English can sometimes be changed into syllogisms by a change in wording (as long as the meaning is not changed):

   - If an argument contains an active verb, the action can be moved into the predicate. E.g. *All boy scouts do good deeds.* becomes *All boy scouts are people who do good deeds.*

   - Quantifiers are sometimes assumed or expressed in non-standard form.

     Dogs are mammals means *All dogs are mammals.*

     Dogs are sometimes vicious means *Some dogs are vicious.*

   - Sometimes people use equivalent terminology rather than repeating a term.

No wealthy people are vagrants.

All attorneys are rich people.

No lawyers are tramps.

- Some arguments are about time or place, but this is not explicit.

  There are plants growing here, and since vegetation requires water, water must be present.

3. Singular propositions can be construed as universal propositions.

   All dogs are mammals.
   Fido is a dog.
   Fido is a mammal.

   This argument is intuitively valid. To show its validity, we treat the minor premise as referring to every member of the class of which Fido is the only member.

   DaM
   $\underline{FaD}$
   FaM

Let's try an example:

Some non-drinkers are athletes, because no drinkers are persons in perfect physical condition, and some people in perfect physical condition are not non-athletes.

$Po\overline{A}$
$\underline{DeP}$
$\overline{D}iA$

On the face of it, this looks like an example of drawing an affirmative conclusion from a negative premise, as well as exclusive premises – a double fallacy. However, we can't apply the rules until we get the argument into standard form.

$Po\overline{A} \leftrightarrow PiA$
$\underline{DeP}$
$\overline{D}iA \leftrightarrow Ai\overline{D} \leftrightarrow AoD$

Converting the conclusion caused the major term to become the minor, and *vice versa*. So we must rewrite the argument:

DeP
<u>PiA</u>
AoD

A Venn diagram test will show that this argument is valid.

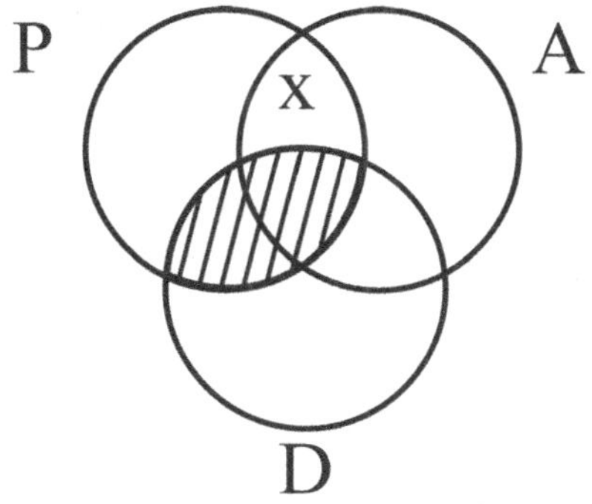

<u>Exercises</u>: For this groups of arguments, you will need to do all the things that were required for the arguments in Part I. In addition, you may need to use synonyms, conversion and obversion, and other creative techniques to put the arguments into standard form. You should also use Venn diagrams to test the validity of the arguments.

1. Some minerals are rare and costly substances, but no jewels are non-minerals; hence some jewels are rare and costly substances.
2. In some African nations the majority of Christians tend to be Anglican, since in all nations colonized by Britain the majority of Christians tend to be Anglican, and some African nations were colonized by Britain.
3. Things that ignite easily are unsafe, so all safe things are non-explosives, since all explosives are things that ignite easily.
4. All of your worldly goods are changeable things, for none

of your worldly goods are immaterial, and no material things are unchangeable things.

5. All those who are neither Democrats nor Republicans are people who are excluded from closed party primaries; therefore no independents are either Democrats or Republicans, for no people who are included in closed party primaries are independents.

6. Every mortal is imperfect, and no humans are immortal, so no humans are perfect.

7. All professional basketball players are no less than six feet tall, since no people that are less than six feet tall are taken seriously by the coaches, and all professional basketball players are people who were taken seriously by the coaches.

8. All syllogisms having two negative premises are invalid. Some valid syllogisms are sound. Therefore some unsound arguments are syllogisms having two negative premises.

9. All syllogisms having two negative premises are invalid. Some valid syllogisms are sound. Therefore some sound arguments are not syllogisms having two negative premises.

10. People who are frequently intoxicated are undependable employees, so all employees who are dependable are non-alcoholics, since all alcoholics are frequently intoxicated.

11. Some lazy people are rich, and rich people are lucky, so some people who are not lazy are unlucky.

12. It must have rained lately, because the fish are not biting, and fish never bite after a rain.

13. The fish won't bite today, because it just rained and the fish never bite after a rain.

14. Eighteen-year-olds are permitted to vote. But not everyone who is permitted to vote actually does vote. So there must be some eighteen-year-olds who do not vote.

15. Since no nonresidents are citizens and all voters are citizens, it must be the case that all voters are residents.
16. Because Harry enjoys himself only when he has lots of money, and because Harry always enjoys going out with Jane, it follows that Harry only goes out with Jane when he has lots of money.

Enthymemes

An argument that omits a premise or conclusion is said to be enthymemic. In ordinary conversation, we frequently omit premises that are regarded as common knowledge, or we leave it to the audience to draw the conclusion if we feel that the conclusion is obvious. Sometimes we omit both a premise and the conclusion.

- Omitted premise: I know the fish won't bite today, because they never bite right after a hard rain.

    Omitted premise: We had a hard rain today.

- Omitted conclusion: The bigger the burger, the better the burger, and burgers are bigger at Burger King.

    Omitted conclusion: Burgers are better at Burger King.

- Both: If Edwards is honest, then I'm a monkey's uncle.

    Omitted premise: I'm not a monkey's uncle.

    Omitted conclusion: Edwards is not honest.

In evaluating an enthymeme, we supply the missing proposition in such a way as to make the argument valid, if possible. In doing this, we will sometimes find that the required premise is implausible, so that the argument, while valid, is not sound.

Example of a valid argument that is not sound: My wife says I'm an alcoholic, but I can't be an alcoholic because I don't go to AA meetings.

AaAA
<u>IeAA</u>
IeA

The supplied major premise is "All alcoholics go to AA meetings", a very questionable premise at best.

Let's look at one that's not valid: Leslie Cole is well groomed, and we know what that implies because all successful people are well groomed. The conclusion we are supposed to draw is that Leslie Cole is successful. However, that would be the fallacy of Undistributed Middle:

SaW
LaW
LaS

Exercises: Complete the following enthymemes so as to make them into valid syllogisms. Indicate whether you have to supply a premise or conclusion. Rewrite the completed argument in standard form. The first one is done for you as an example. (Note that you may have to supply an implausible premise – or one that is known to be false – in order to make some of these arguments valid.)

1. Mary attended the opera, so her lamb must have attended the opera, too.

   Omitted premise: Everywhere that Mary went, the lamb was sure to go.

   In standard form:

   All places that Mary goes are places the lamb also goes.
   The opera is a place that Mary goes.
   Therefore the opera is a place that the lamb also goes.

2. No honest men are crooks. It follows then that no businessmen are honest.

3. From the Sermon on the Mount: "Blessed are the peacemakers, for they shall be called children of God."

4. Surely you didn't put bleach in the colored clothes, because only an idiot would do that.

5. Harry can't buy whiskey since he's not 21.

6. No enthymemes are complete, so this argument is incomplete.
7. No enthymemes are complete, and this argument is an enthymeme, so . . .
8. You can't vote because you haven't registered.
9. People who want to be elected have to respond to their critics. I know you want to be elected. I leave it to you to draw your own conclusion.
10. Andy's a politician, so you can bet he's looking out for himself.
11. This is a difficult course, so not everyone will pass it.
12. This gem can't be a diamond because it won't even scratch glass.
13. A minimum of 70 points is required to pass this exam, Mr. Fluke. You scored 68. You know what that means.
14. The reason that men dread true liberty is that it means responsibility.
15. I know it will rain today because I just washed my car.
16. Interest rates must be down. A lot more houses are for sale.
17. No dogs are cats and some dogs are beagles. What can you infer from these facts?

Sorites

Sometimes an argument is a chain of reasoning containing three or more premises. It may be possible to treat such an argument as a series of enthymemes with all conclusions suppressed except the last one. If so, the argument is a sorites.

Consider the following example:

> All musicians are entertainers.
> No rocket scientists are entertainers.
> Some lead singers are bass players.

All bass players are musicians.
Therefore some lead singers are not rocket scientists.

We can treat this as three syllogisms as follows:

All musicians are entertainers.
No rocket scientists are entertainers.
No rocket scientists are musicians.

No rocket scientists are musicians.
All bass players are musicians.
No bass players are rocket scientists.

No bass players are rocket scientists.
Some lead singers are bass players.
Therefore some lead singers are not rocket scientists.

All three arguments are valid; therefore, the sorites is valid.

To evaluate a sorites

1. Symbolize it so that each term occurs twice (if possible), performing any necessary obversion, conversion, etc.

   Example:

No skiers are non-athletic.	$Se\overline{A} \leftrightarrow SaA$
Some nutritionists are skiers.	NiS
Athletes are not brawny.	AeB
Some nutritionists are not brawny.	NoB

2. Find 2 premises with a common term. Supply a conclusion so as to make this syllogism valid, if possible.

   SaA
   NiS
   NiA

3. Repeat step 2, using the supplied conclusion as one of the premises.

   NiA
   AeB
   NoB

4. The final conclusion should be equivalent to the conclusion of the sorites.

To see how this works with an invalid sorites, consider the following example:

> No letter carriers are civil servants. Everyone who is looking for a job is unemployed. Some letter carriers are unemployed. Therefore there are some people looking for jobs who are not civil servants.

1. First, symbolize the argument (and get rid of extra terms, if necessary):

   LeC
   JaU
   LiU
   JoC

2. Next, break the sorites into its constituent syllogisms:

LeC	UoC
LiU	JaU
UoC	JoC

3. But the second syllogism commits the fallacy of Undistributed Middle, so the sorites is not valid.

Exercises: Evaluate the following sorites by breaking them down into their constituent syllogisms. Identify any fallacy committed.

1. I know that Eunice, the small town in which I live, is in the United States, because Eunice is in St. Landry Parish, and St. Landry Parish is in Louisiana, and Louisiana is in the United States.

2. Whenever it rains the streets get wet, and whenever the streets get wet there are more accidents. On days that there are more accidents, the police department has much more work to do. So the police department has much more work to do on rainy days.

3. All murderers are sinners and all sinners will be punished in the next life. But Boudreaux is not a murderer, so he will not be punished in the next life.

4. Logic classes are always difficult. Difficult classes are always poorly attended. Sometimes poorly-attended classes are canceled. So logic classes are sometimes canceled.

5. All musicians are entertainers. No plumbers are entertainers. Some opera stars are wealthy people. All opera stars are musicians. Therefore some plumbers are not wealthy people.

6. Everyone who is looking for a job is unemployed. All doctors are highly educated, but some are unemployed. Therefore there are some people looking for jobs who are highly educated.

7. Anything that promotes skin cancer causes death. Whatever preserves the ozone layer prevents the release of chlorofluorocarbons. Nothing that resists skin cancer increases ultraviolet radiation. Anything that destroys the ozone layer increases ultraviolet radiation. There are packaging materials that release chlorofluorocarbons. Nothing that causes death should be legal. Therefore some packaging materials should be illegal.

Enthymemic Sorites

A sorites with its conclusion suppressed is called an enthymemic sorites. To evaluate such an argument, we combine the techniques used in evaluating enthymemes and sorites.

Consider the following example: "All flying saucers have ailerons. Anything with ailerons can turn on its longitudinal axis. Anything that can turn on its longitudinal axis possesses aesthetic value. Cockroaches never possess aesthetic value." From this we may validly conclude that cockroaches are not flying saucers, as the following breakdown shows: (Note: The solution of this sorites is my own, but the sorites came from another logic book, as did the one in #7 above. I would like to acknowledge the authors, but unfortunately I have lost the references.)

AaL	LaE	FaE
FaA	FaL	CeE
FaL	FaE	CeF

Exercises: In each of the following enthymemic sorites, supply the conclusion so as to make the argument valid (if possible).

1. Fido is a dog. Dogs are mammals. Mammals are animals. Animals are living things.

2. Texans are ranchers and some ranchers are shepherds. All shepherds need dogs to help them.

3. Logic courses are difficult. Courses that are difficult seldom have a large enrollment. Courses with low enrollment are always cancelled by the Dean.

4. The following sorites is valid but its conclusion is missing. You need to prove its validity by decomposing it into its constituent syllogisms. In the process of doing this, you should produce the missing conclusion. (Hint: the conclusion is a popular topic among many students.)

   $\overline{G}a\overline{U}$
   $Se\overline{R}$
   XeH
   RaU
   GaH

Modern Logic

Mathematical Logic

In mathematics, an operation such as addition is indicated by the expression 3+5. In that expression, + is an operator, and the numbers 3 and 5 are called operands. If we replace the numbers in a numeric expression with variables – such as X+Y – it would be clear that the letters represented numbers.

If we construct an expression containing more than one operator, we can have ambiguity of meaning. For example, the expression 3+5×2 could yield either 16 or 13, depending on whether the addition or the multiplication is done first. One way to resolve the ambiguity is to use parentheses to show which operation should be performed first:

$(3+5) \times 2 = 8\times2 = 16$

$3 + (5\times2) = 3+10 = 13$

Logical Operators

We can represent logical operations by the use of logical operators. For example, the operation of negation is represented by NOT. In numeric operations the operands are numbers, and variables can be used to represent numbers. In logical operations the operands are truth-values, and variables can be used to represent statements that can be either true or false. Thus, if A is a statement, the negation of that statement would be represented by NOT A.

Besides negation, the fundamental logical operations are conjunction and disjunction, represented by the logical operators AND and OR. It makes sense to negate a single statement, but you can't conjoin (or disjoin) one thing. In other words, the operations of conjunction and disjunction require two operands: A AND B or A OR B.

If we wish to construct complex expressions, such as A AND B OR C we will use parentheses to avoid ambiguity: (A AND B) OR C or A AND (B OR C)

Truth Tables

Truth tables can be used to show all possible truth-values for an expression. For example, there are only two possible values for the expression NOT A.

A	NOT A
T	F
F	T

With two-operand expressions, there are four possible combinations of truth-values:

A	B	A AND B
T	T	T
T	F	F
F	T	F
F	F	F

Truth tables can also be regarded as defining how an operator is to be used. For example, the word "or" is ambiguous in English. It can mean either/or (exclusive) or and/or (inclusive). To avoid ambiguity, the logical OR is given the inclusive meaning:

A	B	A OR B
T	T	T
T	F	T
F	T	T
F	F	F

If we wished to define an operator for the exclusive use of "or" (XOR, as it is sometimes called), we can construct a complex expression. In ordinary English, we say that either/or means one or the other but not both. The corresponding logical expression is (A OR B) AND NOT (A AND B). The truth table would look like this:

A	B	A OR B	A AND B	NOT (A AND B)	(A OR B) AND NOT (A AND B)
T	T	T	T	F	F
T	F	T	F	T	T
F	T	T	F	T	T
F	F	F	F	T	F

The truth table verifies our everyday understanding of the exclusive "or"; that is, such a statement is true only when the constituent statements have different truth-values. If it's truly an either/or situation, they can't both be true at the same time. You can vote for a Democrat or a Republican for President, but not both. (Perhaps it is more accurate to say that you cannot legally vote for both. However, I am reminded of the old Louisiana election-day maxim: "Vote early and vote often!").

We have seen that a truth table with only one variable has two rows, while one with two variables requires four rows. If we add a third variable, the number of rows necessary to show all possible combination of truth-values would be eight. In general, the number of rows in a truth table is 2^N, where N is the number of variables.

We can also use truth tables to show the difference that parentheses make. For example, the expression A OR B AND C is ambiguous; it could mean either A OR (B AND C) or (A OR B) AND C.

A	B	C	A OR B	(B AND C)	A OR (B AND C)	(A OR B) AND C
T	T	T	T	T	T	T
T	T	F	T	F	T	F
T	F	T	T	F	T	T
T	F	F	T	F	T	F
F	T	T	T	T	T	T
F	T	F	T	F	F	F
F	F	T	F	F	F	F
F	F	F	F	F	F	F

The truth table shows that the two statements have different meanings, because their truth-values are different in the second and fourth rows.

Exercises: Construct a truth table for each of the following expressions.

1. A OR (A AND B)
2. A AND NOT (A OR B)
3. A OR (B AND C)
4. (A AND NOT B) OR (B AND C)
5. NOT [A AND NOT (A OR NOT B)]

Mathematical Symbolism

In modern logic, it is customary to use mathematical symbols instead of words for operators. This list includes the three primitive operators plus two additional ones.

Word	Symbol	Name
NOT	~	tilde
AND	•	dot
OR	v	vel
IMP	→	implies
EQUIV	↔	equivalent

These last two operators are derived from the primitives. The → operator is used to symbolize if/then (conditional) statements. However, it does not capture the full meaning of conditionals. It is better regarded as a minimal statement: if A implies B, what is the least that must be true in all cases? It is that it is impossible for A to be true and B to be false. Thus A→B is said to mean the same as ~(A•~B). We can show this derived meaning with a truth table:

A	B	~B	A•~B	~(A•~B)	A → B
T	T	F	F	T	T
T	F	T	T	F	F
F	T	F	F	T	T
F	F	T	F	T	T

Suppose Lacey Underalls is struggling with logic: she failed the mid-term and hasn't done well on the homework. So she goes to Prof. Meaney and asks him what it will take for her to get a C in the course. He says, "If you make at least 85 on the final, I'll give you a C for the course." If she gets a C in the course, does that mean she made at least 85 on the final? Not necessarily. The teacher may have been more generous than he promised to be. But if she receives a D or F in the course, we can infer one of two things: either that she made less than 85 on the final, or that Prof. Meaney reneged on his promise. In general, the only time we can be sure that a conditional statement is false is when the antecedent is true and the consequent is false. That corresponds to row 2 in the truth table.

Now let's look at equivalence. What do we mean when we say that two statements, A and B, are equivalent to each other? One thing we mean is that A implies B and B implies A: (A→B) • (B→A). Alternatively, we could say that if two statements are equivalent, either they're both true of they're both false: (A•B) v (~A•~B). We can use a truth table to show that these two expressions have the same truth-value:

1	2	3	4	5	6	7	8			
A	B	~A	~B	A•B	~A•~B	A→B	B→A	7•8	5v6	A↔B
T	T	F	F	T	F	T	T	T	T	T
T	F	F	T	F	F	F	T	F	F	F
F	T	T	F	F	F	T	F	F	F	F
F	F	T	T	F	T	T	T	T	T	T

The last column contains the definition of the equivalence operator. A and B are not equivalent unless they have the same truth-value, so the statement that they are equivalent (A ↔ B) is false if their truth-values are different.

Exercises: Construct a truth table for each of the following expressions.

1. A • (AvB)
2. A • (BvC)
3. [(A → B)•A] → B
4. ~(A → B) • (~AvB)
5. ~(A → B) • (~BvC)
6. ~(A → B) ↔ (~AvB)

Summary of Logical Operators

	Traditional Notation			Mathematical Notation		
NOT	X	NOT X		X	~X	
	T	F		T	F	
	F	T		F	T	
AND	X	Y	X AND Y	X	Y	X•Y
	T	T	T	T	T	T
	T	F	F	T	F	F
	F	T	F	F	T	F
	F	F	F	F	F	F
OR	X	Y	X OR Y	X	Y	XvY
	T	T	T	T	T	T
	T	F	T	T	F	T
	F	T	T	F	T	T
	F	F	F	F	F	F

	Traditional Notation			Mathematical Notation		
IMP	X	Y	X IMP Y	X	Y	$X \rightarrow Y$
	T	T	T	T	T	T
	T	F	F	T	F	F
	F	T	T	F	T	T
	F	F	T	F	F	T
EQV	X	Y	X EQV Y	X	Y	$X \leftrightarrow Y$
	T	T	T	T	T	T
	T	F	F	T	F	F
	F	T	F	F	T	F
	F	F	T	F	F	T

Translation

There is no mechanical technique for translating from ordinary English into mathematical notation, but here are some helpful hints:

1. Try to determine the principal operation: is the statement a conjunction, an implication, etc.?
2. Be careful about the scope of **not**. Note the difference between:

   Art will go but Betsy won't: $A \bullet \sim B$

   Art and Betsy won't both go: $\sim(A \bullet B)$

3. Do not expect a strict correlation between sentence form and logical type. Note that the following propositions are all correctly symbolized as $C \rightarrow S$.

   If you're changing to daylight savings time, you need to set your clock up an hour.

   You need to set your clock up an hour if you're changing to daylight savings time.

Changing to daylight savings time means that you need to set your clock up an hour.

4. Do not expect the physical order of the sentence to determine the physical order of operators and operands. Note the two different, but equivalent, ways to symbolize the following statement. (There are truth tables on the next two pages showing that the two statements have the same truth-values.)

   If Betsy learned to read and write well, she could get a good job and make lots of money – if she didn't have eight children.

   $(L \bullet {\sim}E) \rightarrow (J \bullet M)$ $\quad {\sim}E \rightarrow (L \rightarrow (J \bullet M))$

5. Remember that the symbols of mathematical logic cannot capture all the meaning of ordinary words. Thus, all conjunctions – and, but, although, etc. – are indifferently symbolized by •.

6. Universal categorical propositions are treated as hypotheticals. Thus:

   - SaP becomes $S \rightarrow P$.

   Example: "All Democrats are liberals" becomes "If you're a Democrat, then you're a liberal".

   - SeP becomes $S \rightarrow {\sim}P$.

   Example: "No Republicans are liberals" becomes "If you're a Republican, then you're not a liberal".

7. Particular categorical propositions are treated as the negations of their contradictories. Thus:

   - SiP becomes ${\sim}(S \rightarrow {\sim}P)$.

   Example: "Some Republicans are liberals" becomes "It's not true that if you're a Republican, then you're not a liberal".

   - SoP becomes ${\sim}(S \rightarrow P)$.

   Example: "Some Democrats are not liberals" becomes "It's not true that if you're a Democrat, then you're a liberal".

Here are the truth tables for the two expressions in item 4 above. Note that the word ALL in both tables refers to the full expression. Note also that the truth-values for ALL are the same in both truth tables.

(L•~E) → (J•M)

E	J	L	M	~E	(L•~E)	(J•M)	ALL
T	T	T	T	F	F	T	T
T	T	T	F	F	F	F	T
T	T	F	T	F	F	T	T
T	T	F	F	F	F	F	T
T	F	T	T	F	F	F	T
T	F	T	F	F	F	F	T
T	F	F	T	F	F	F	T
T	F	F	F	F	F	F	T
F	T	T	T	T	T	T	T
F	T	T	F	T	T	F	F
F	T	F	T	T	F	T	T
F	T	F	F	T	F	F	T
F	F	T	T	T	T	F	F
F	F	T	F	T	T	F	F
F	F	F	T	T	F	F	T
F	F	F	F	T	F	F	T

~E → (L→(J•M))

E	J	L	M	~E	(J•M)	L→(J•M)	ALL
T	T	T	T	F	T	T	T
T	T	T	F	F	F	F	T
T	T	F	T	F	T	T	T
T	T	F	F	F	F	T	T
T	F	T	T	F	F	F	T
T	F	T	F	F	F	F	T
T	F	F	T	F	F	T	T
T	F	F	F	F	F	T	T
F	T	T	T	T	T	T	T
F	T	T	F	T	F	F	F
F	T	F	T	T	T	T	T
F	T	F	F	T	F	T	T
F	F	T	T	T	F	F	F
F	F	T	F	T	F	F	F
F	F	F	T	T	F	T	T
F	F	F	F	T	F	T	T

Symbolic Translation Exercises: Translate the following sentences into the symbolism of modern logic, using the indicated symbols.

Group 1

R: Sen. Jones will be re-elected.

T: He raises taxes.

S: He eliminates wasteful government spending.

1. Sen. Jones will be re-elected if he can eliminate wasteful government spending and doesn't raise taxes.

2. Either Sen. Jones eliminates wasteful government spending without raising taxes, or he won't be re-elected.
3. If Sen. Jones eliminates wasteful government spending but also raises taxes, he won't be re-elected.

<u>Group 2</u>

A: Trump is a pompous ass.

R: Trump will be re-elected.

P: Trump makes (will make) an effective populous appeal.

4. Trump is a pompous ass, but he will be re-elected.
5. If Trump is not able to make an effective populous appeal, he will not be re-elected.
6. Even if Trump makes an effective populous appeal, he will not be re-elected because he is a pompous ass.

<u>Group 3</u>

G: Germany mobilizes.

F: France protests (will protest) to the UN.

I: Italy will call for a meeting of the European Union.

7. If Germany mobilizes or France protests to the UN, then Italy will call for a meeting of the European Union.
8. If Germany mobilizes, then if France protests to the UN, then Italy will call for a meeting of the European Union.
9. Germany will mobilize if either France protests to the UN or Italy calls for a meeting of the European Union.
10. France will protest to the UN unless Italy calls for a meeting of the European Union.

<u>More Symbolic Translation Exercises</u>: Choose the best translation for each of the following sentences, using the indicated symbols. Circle the letter corresponding to your choice.

1. "No easy exercise ever gives me a headache." E: these are easy exercises. H: these exercises give me a headache.

   (a) EvH

   (b) E•H

   (c) ~H → E

   (d) ~E → H

   (e) E → ~H

2. "If I don't pay the butcher he won't let me have the roast I want for dinner, and without the roast I can't prepare the good meal I had planned." P: I prepare the good meal I had planned. B: I pay the butcher. R: I get the roast I want for dinner.

   (a) (~B →~R) • ~P

   (b) ~B → (~R →~P)

   (c) (~B →~R) • (~R →~P)

   (d) (~B•~R) → (~R•~P)

   (e) (~Bv~R) → (~Rv~P)

3. "If you are not good in math, then you won't do well in logic and you won't be able to satisfy the college's critical reasoning requirement." G: You are good in math. L: You do well in logic. S: You will be able to satisfy the college's critical reasoning requirement.

   (a) ~(G →L)•~S

   (b) ~G → (~L•~S)

   (c) ~[G → ~(L•S)]

   (d) ~[G → (LvS)]

4. "Either bourbon or Scotch is your favorite whiskey, but not both." B: Bourbon is your favorite whiskey.
   S: Scotch is your favorite whiskey.

   (a) (BvS) •~(B•S)

(b) (BvS) • (~B•~S)

(c) (BvS) → (~B•~S)

(d) (B•S) v~(BvS)

5. "Germany will mobilize its troops and either France will declare an emergency or Italy will appeal to the European Union." G: Germany will mobilize its troops. F: France will declare an emergency. I: Italy will appeal to the European Union.

   (a) (GvF) • I

   (b) G v (F•I)

   (c) G • (FvI)

   (d) (G•F) v I

   (e) G → (FvI)

Still More Symbolic Translation Exercises: Translate each of the following statements into the symbolism of modem logic.

1. Joe is a fool, but his wife is smart and level-headed.
2. Calculus is required for computer science majors but not for English majors.
3. It is not true that Sen. Jones was both corrupt and dishonest.
4. Sen. Jones will be re-elected if he eliminates wasteful government spending but doesn't raise taxes.
5. No sinners go to heaven.
6. Blessed are the peacemakers.
7. No enthymemes are complete.
8. If enough constituents complain about Sen. Jones, the FBI will launch an investigation of him.
9. Either his constituents will complain about Sen. Jones or they won't.

10. If we raise the entrance standards, some of our students will be better prepared for college, but some won't.

11. If the U.S. doesn't eliminate its nuclear arsenal, Russia won't; and if Russia won't, the U.S. won't; and so there will be a nuclear catastrophe.

12. Neither the U.S. nor Russia will eliminate its nuclear arsenal.

Computer Logic

Circuit Design

Computers routinely perform logical operations. For example, a payroll program determines which employees are entitled to the overtime premium using code such as the following:

If Hours > 40 then GrossWages =
40 * PayRate + (Hours – 40) * PayRate * 1.5

Consider the difference between the two expressions in the code:

Hours > 40
40 * PayRate + (Hours – 40) * PayRate * 1.5

When we have values for the variables, evaluating the second expression will produce a number, but evaluating the first expression produces a truth value (either true or false). Thus, calculating the gross pay for employees who are entitled to the overtime premium is an exercise in arithmetic, but determining who those employees are, is a matter of logic.

To be able to perform logical operations, a computer has logic circuits built into its circuit board. The fundamental logic circuits implement the logical operators NOT, AND, and OR.

A simplified form of the AND circuit looks something like this for the expression A AND B:

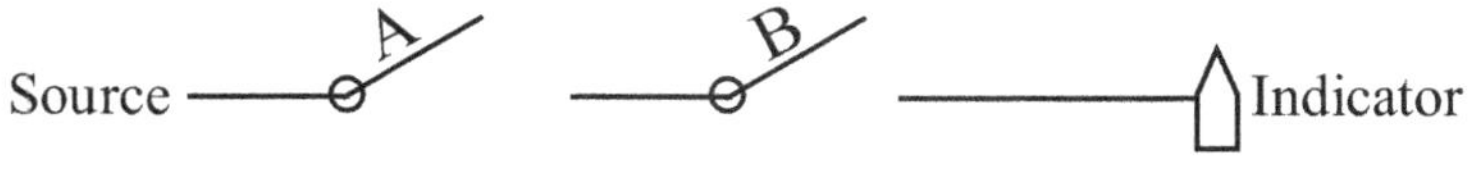

This assumes a flow of electricity from the source to the light (or some other indicator). To complete the circuit, both the A and B gates must be closed.

Before drawing the OR circuit, we need to discuss the ambiguity of the word *or* in English: it can mean either/or (exclusive) or and/or (inclusive). By definition, the OR operator is defined as the inclusive *or*. The circuit for the expression A OR B would look something like this:

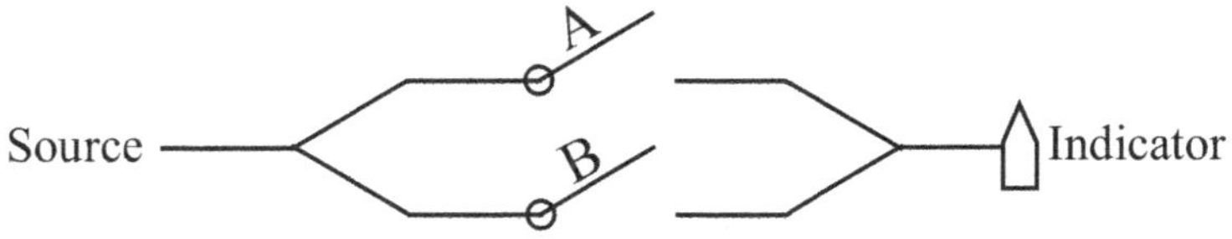

To complete this circuit, either the A gate or the B gate, or both, may be closed.

Although the NOT operator is the simplest of the three basic operators, its circuit is a bit more complicated.

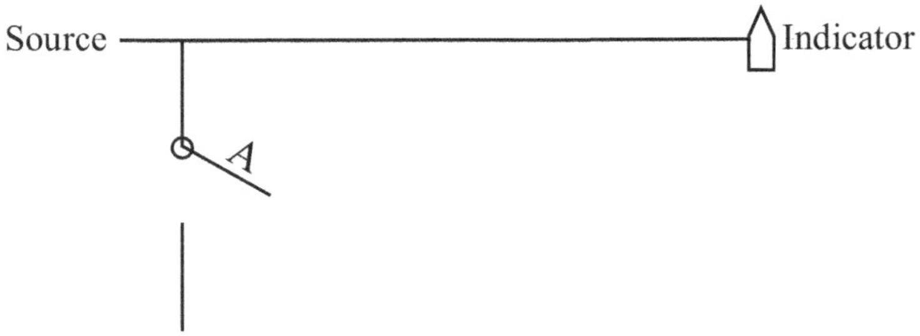

This circuit works because electricity will follow the shortest path or the path of least resistance. So if the gate is closed as it would be for the expression NOT A, electricity will never get to the indicator. If A is true, then NOT A will be false, in which case the light will not come on.

The basic circuits are called gates and are symbolized as follows:

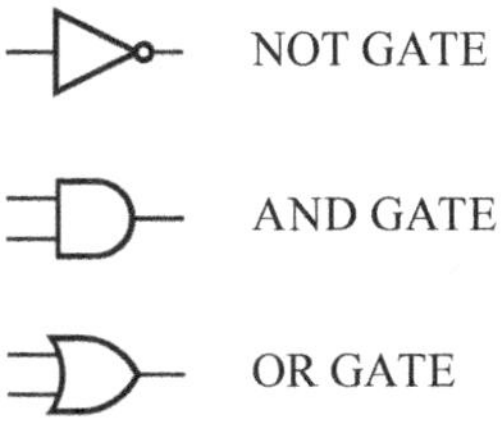

Input Output

Now let us return to the OR operator. It is defined as the inclusive *or*. If we wish to use an exclusive *or* (called XOR), we require a derived circuit. What is meant by A XOR B is that A and B cannot both be true. There are two ways to formulate the expression:

(A AND NOT B) OR (NOT A AND B)
in mathematical notation: (A•~B) v (~A•B)

(A OR B) AND NOT (A AND B)
in mathematical notation: (AvB) •~(A•B)

The equivalence of these two expressions can be shown by the following truth table:

A	B	~A	~B	A•~B	~A•B	(A•~B) v (~A•B)	A v B	A•B	~(A•B)	(A v B)•~(A•B)
T	T	F	F	F	F	F	T	T	F	F
T	F	F	T	T	F	T	T	F	T	T
F	T	T	F	F	T	T	T	F	T	T
F	F	T	T	F	F	F	F	F	T	F

Since the truth values are the same for the two expressions, they are logically equivalent. Since they are logically equivalent, we can design a circuit for either expression. Here is the circuit for the first one.

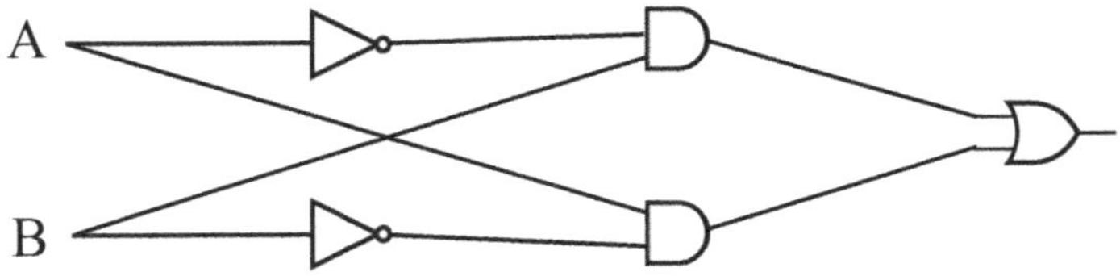

CIRCUIT FOR XOR GATE

Boolean Algebra

The term "Boolean algebra" reflects the developer, George Boole. Some of the fundamental laws of Boolean algebra are the laws of commutation, association, distribution, and DeMorgan's theorem. Each of these has two forms:

- Commutation:

  AvB = BvA

  A•B = B•A

- Association:

  Av(BvC) = (AvB)vC

  A•(B•C) = (A•B)•C

- DeMorgan's Theorem:

  ~(A•B) = ~Av~B (NAND)

  ~(AvB) = ~A•~B (NOR)

- Distribution:

  Av(B•C) = (AvB)•(AvC)

  A•(BvC) = (A•B)v(A•C)

The terms NAND and NOR are obviously short-hand for NOT AND and NOT OR. Here are their circuits:

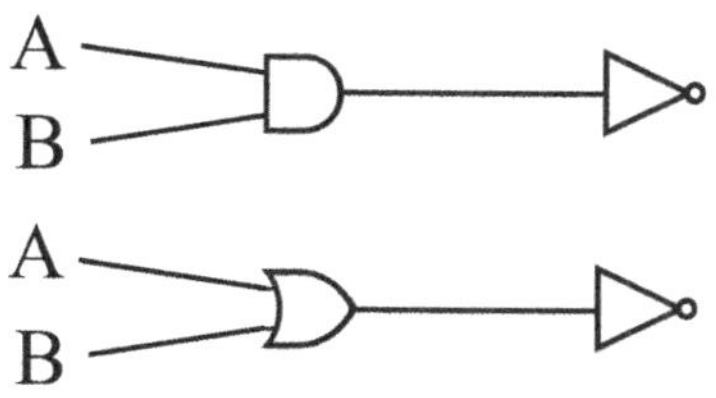

The following truth-table shows the equivalency for the first form of DeMorgan's theorem (the NAND expression):

A	B	A•B	~(A•B)	~A	~B	~Av~B
T	T	T	F	F	F	F
T	F	F	T	F	T	T
F	T	F	T	T	F	T
F	F	F	T	T	T	T

Logic circuits have useful real-world applications, such as in the design of burglar alarms. Suppose a restaurant has a front door (F) and back door (B), either of which can be open when the restaurant is open for business (O). If the restaurant is not open, the alarm should sound if either door is open. The expression for the alarm is: ~O•(FvB)

Exercises

1. Construct a truth-table for the burglar alarm expression.
2. Construct a truth-table to show the equivalence of either form of the law of distribution.
3. Draw the circuit for the burglar alarm.
4. Draw the circuit for the following expression. ~(A•B) v~C
5. Write the expression that corresponds to the following circuit.

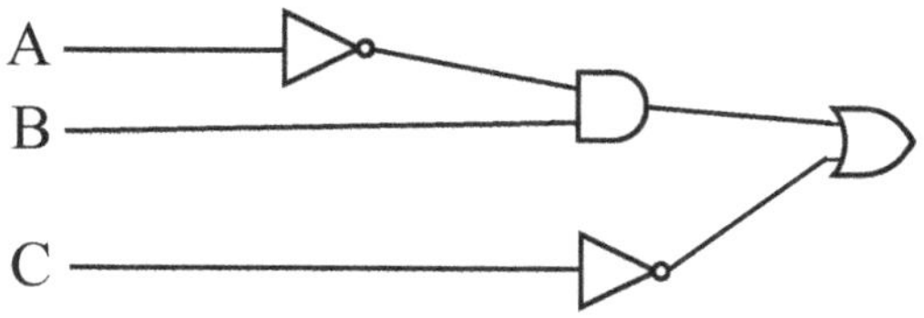

6. Explain why this circuit is called the Hamlet circuit.

7. Boudreaux's Bakery has a front door and a back door, both of which may be open when the bakery is open for business, but neither of which should be open when the bakery is closed. Boudreaux wants a special burglar alarm for his business that he doesn't have to set, but that will be automatically set when the front door is locked. However, he knows that there are times when the staff is there after hours working on special orders and they want the back door open for ventilation. Also, there are times when the bakery is closed but he receives merchandise after hours and the front door has to be opened. So he needs an over-ride switch that he can use to turn the alarm off on special occasions. Draw a circuit for the burglar alarm. It will have two inputs: the lock on the front door and the over-ride switch, plus the front and back doors. Assume that L represents the door being locked, F and B refer to the front or back door being open, and O designates the over-ride switch, the expression for the circuit would be [L•(FvB)]•~O.

Decision Structures

One of the things required of computer programmers is to implement decision structures in their computer code. For this to be done correctly, the programming specifications must be very accurate. There are many ways to correctly specify decision structures. We will look at two: decision trees and flowcharts. These tools are by no means limited to computer programmers. For example, the Internal Revenue service will often use flowcharts to explain the details of the tax code.

Let us take a hypothetical example of some complexity. Suppose Waxahachie Widget Works employs three categories of workers: administrators, who receive a straight salary, sales personnel, who receive a base salary plus a commission based on their sales, and hourly workers. Suppose moreover that the sales commission rate is variable: 4% of the first $5,000, 5% of the next $2,500, and 6% of the amount over $7,500. For example, if Bubba Boudreaux's sales are $8,000, he will receive $355 in commission: 4% of $5,000 plus 5% of $2,500, plus 6% of $500.

To add a bit more complexity, let us suppose that the overtime premium is variable. Normally it will be time-and-a-half for hours over 40, but if any of that overtime occurs on a holiday, the overtime premium will be double time.

Note that the double-pay premium only applies to overtime hours. If the holiday hours are part of the employee's 40-hour work week, the employee will receive time-and-a-half for the holiday hours. Note also that an employee may have both regular overtime hours, for which the employee will receive time-and-a-half, and holiday overtime hours at double pay.

To be able to write the program specs, we need some way to identify the employee type. Let us assume a numeric code is used as follows: 1 for administrators, 2 for sales personnel, and 3 for hourly workers. Suppose moreover that the employee time cards contain a breakdown of hours into regular hours (RegHours) and holiday hours (HolHours).

The scenario I have described is not as complex as some of the IRS code, but it should be sufficiently complex to illustrate both decision trees and flowcharts.

In drawing a decision tree, we normally have several columns labeled with various factors that must be considered in determining the output, plus a final column for the output. In this example, our output will be gross pay. The factors that must be considered are the employee code (EmpCode), sales, regular hours (RegHours) and holiday hours (HolHours).

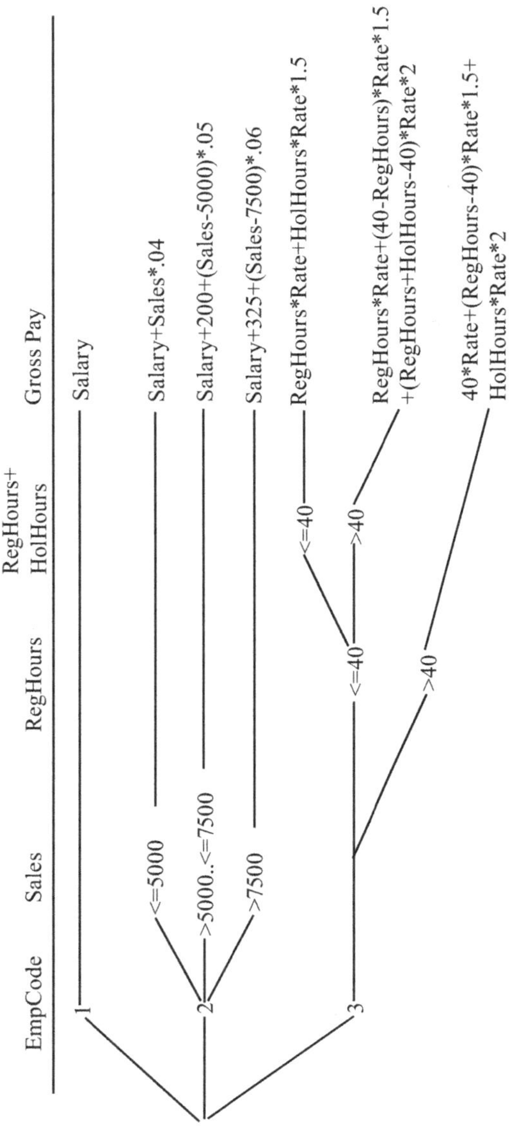

People who draw flowcharts for programmers often use different shapes for different kinds of operations, such as input and processing. This complication is unnecessary for our purposes. All we really need are circles and arrows, a place to start, and a specification of the outcome (gross wages) for each category of employee.

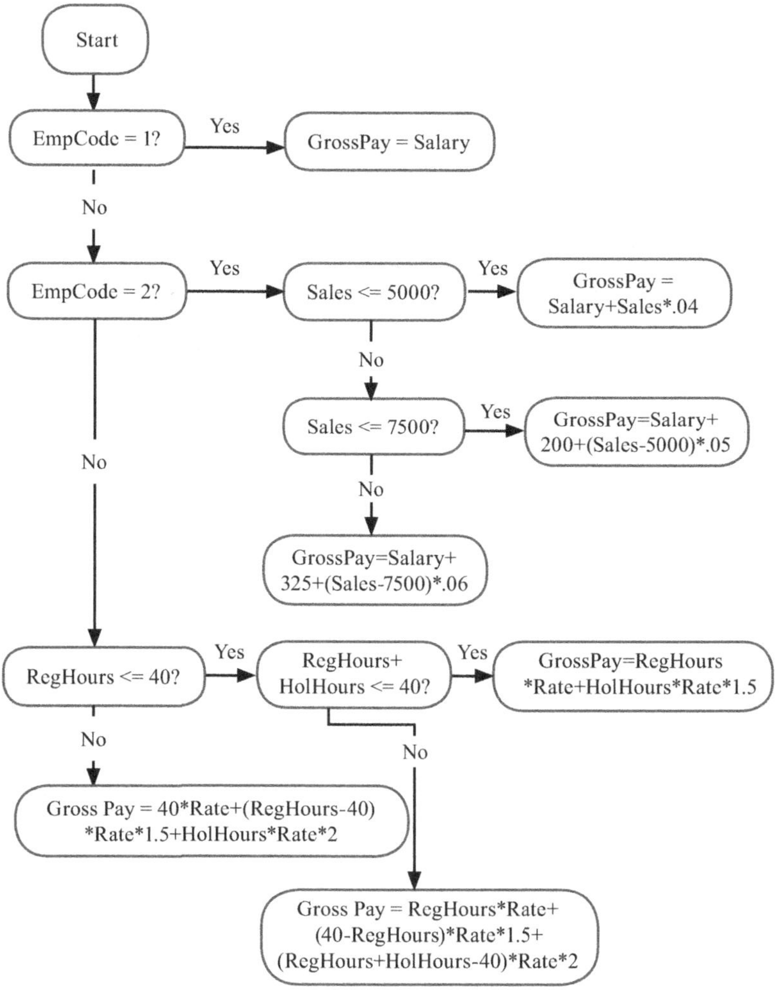

<u>Exercises</u>

1. Suppose that the management of Waxahachie Widget Works decides that the business will no longer be open on holidays, so it will no longer be recording holiday hours for its hourly employees. Your project is to modify the decision tree on the preceding page to reflect this change. The only input for hourly employees will be Hours.

2. Modify the flowchart above to reflect the change described in exercise 1.

Probability

In everyday speech, we use the word "probably" without much precision. If we say that something will probably happen, we mean that we think it's more likely to happen than not. However, there are times when people assign a numeric probability to an event. Perhaps the weatherman will say, "There's a 70% chance of rain along the Louisiana coastline tomorrow." Or maybe a life insurance salesman will say, "According to our actuary tables, a 50-year-old woman in the United States is 3 times as likely as a 50-year-old man to live 25 additional years."

We have probably heard the old saying that 75% of statistics are made up on the spot. And no doubt some people attempt to give their beliefs a specious credibility by claiming precision when none is warranted. But if our intentions are more honorable, we need to first decide what the word "probability" really means. There are at least three possible meanings:

- Subjective expectation: Such a use of "probability" is a statement about the strength of someone's belief. But as we have all discovered from time to time, beliefs and reality don't always coincide. This is well illustrated by the gambler's fallacy, which exists in two different forms. The first form is the expectation that "if you're hot, you're hot"; that is, if you are having a winning streak, it will continue indefinitely. The other form, which is almost the opposite of this one, is that "your number's bound to come up"; that is, if you have had a losing streak, your luck will change if you play just a little longer.
- Empirical evidence: This use of "probability" is based on the record of history. This is presumably the basis of actuary tables. Insurance companies collect data on health, lifestyle, occupational hazards, and other factors related to longevity, and use this information to set life insurance premiums.
- Relative frequency: This is the mathematician's use of "probability". It is based on a simple calculation of odds,

assuming no outside interference. For example, if you roll a die, your chance of rolling 6 is one out of six, assuming that the die is not loaded and that there is no attempt to influence the outcome.

There is no accurate way to measure subjective expectation. In our discussion of probability, we will look at objective ways to calculate the likelihood that an event will occur (or won't occur, as the case may be). Notice the use of the word "calculate". Determining the probability of an event is rarely as simple as the example of rolling 6 with one die. If you're playing craps or poker and the stakes are high, there will be many factors that have to be taken into consideration before you place your bet. The rules that take all these factors into account are collectively known as the probability calculus.

The Probability Calculus

The rules of the probability calculus can be stated using the mathematical symbolism that was introduced in a previous chapter. The only additional thing that requires explanation is the dual use of the letter p : capital P is used to indicate probability, and little p is used to indicate any event. Thus, in the first rule, the expression P(p•~p) can be read "The probability that an event will occur and that the same event will not occur".

General rules of the probability calculus

1. P(p•~p) = 0. This is impossible.
2. P(pv~p) = 1. This is a sure thing.
3. The range of probability for all other events (that is, things that are possible but not certain) is between 0 and 1.
4. P(~p) = 1 - P(p). In other words, the probability that any event won't occur is 1 minus the probability that it will occur.

Conjunction Rules

1. Conjunction of independent events: P(p•q) = P(p) × P(q). To see how this rule is derived, consider the following examples.

   - What are the odds of rolling double sixes? To answer the question, we can enumerate all possible occurrences as follows:

1 , 1	2 , 1	3 , 1	4 , 1	5 , 1	6 , 1
1 , 2	2 , 2	3 , 2	4 , 2	5 , 2	6 , 2
1 , 3	2 , 3	3 , 3	4 , 3	5 , 3	6 , 3
1 , 4	2 , 4	3 , 4	4 , 4	5 , 4	6 , 4
1 , 5	2 , 5	3 , 5	4 , 5	5 , 5	6 , 5
1 , 6	2 , 6	3 , 6	4 , 6	5 , 6	6 , 6

Of 36 possible combinations, 1 is favorable, so the odds are 1 in 36. This is equivalent to the probability of rolling one six, times the probability of rolling another six: $\frac{1}{6} \times \frac{1}{6} = \frac{1}{36}$.

   - What are the odds of tossing heads 3 times in a row? To answer the question, we can enumerate all possible occurrences as follows:

     H H H
     H H T
     H T H
     H T T
     T H H
     T H T
     T T H
     T T T

     Of 8 possible combinations, 1 is favorable, so the odds are 1 in 8. This is equivalent to the probability of tossing one head times the probability of tossing another head times the probability of tossing a third head: $\frac{1}{2} \times \frac{1}{2} \times \frac{1}{2} = \frac{1}{8}$.

2. Conjunction of dependent events: P(p•q) = P(p) × P(q, given p). To see how this rule is derived and what it means, consider the following examples.

   - What are the odds of drawing an ace and then a king from a group of 5 cards: A, K, Q, J, and 10? To answer the question, we can enumerate all possible occurrences as follows:

A , K	K , A	Q , A	J , A	10 , A
A , Q	K , Q	Q , K	J , K	10 , K
A , J	K , J	Q , J	J , Q	10 , Q
A , 10	K , 10	Q , 10	J , 10	10 , J

     Of 20 possible combinations, 1 is favorable, so the odds are 1 in 20. This is equivalent to the probability of drawing an ace from 5 cards times the probability of drawing a king from the remaining 4 cards: $\frac{1}{5} \times \frac{1}{4} = \frac{1}{20}$.

   - What are the odds of selecting 2 red balls (in two draws) from a jar containing 2 red balls and 1 white one? To answer the question, we can enumerate all possible occurrences as follows:

     R R W or R W R or W R R

     Of 3 possible combinations, 1 is favorable, so the odds are 1 in 3. This is equivalent to the probability of drawing 1 red ball out of 3 balls (2 of which are red) times the probability of drawing another red ball from the remaining 2 balls: $\frac{2}{3} \times \frac{1}{2} = \frac{1}{3}$.

Disjunction Rules

1. Disjunction of independent events: P(pvq) = P(p) + P(q). To see how this rule is derived, consider the following examples.

   - What are the odds of rolling a 5 or a 6 on one roll of a die? To answer the question, we can enumerate all possible occurrences as follows:

1 2 3 4 5 6

Of 6 possible occurrences, 2 are favorable, so the odds are 2 in 6 or 1 in 3. This is equivalent to the probability of rolling a 6 plus the probability of rolling a 5: $\frac{1}{6} + \frac{1}{6} = \frac{1}{3}$.

2. What are the odds of tossing 2 heads or 2 tails on 2 tosses of a coin? To answer the question, we can enumerate all possible occurrences as follows:

   H H or H T or T H or T T

   Of 4 possible occurrences, 2 are favorable, so the odds are 2 out of 4 or 1 out of 2. This is equivalent to the probability of tossing 2 heads plus the probability of tossing 2 tails: $\frac{1}{4} + \frac{1}{4} = \frac{1}{2}$.

3. Disjunction of dependent events: P(pvq) = P(p) + P(q) - P(p•q). To see how this rule is derived, consider the following examples.

   - What are the odds of getting at least one 6 out of two rolls of a die? To answer the question, we can enumerate all possible occurrences as follows:

1 , 1	2 , 1	3 , 1	4 , 1	5 , 1	6 , 1
1 , 2	2 , 2	3 , 2	4 , 2	5 , 2	6 , 2
1 , 3	2 , 3	3 , 3	4 , 3	5 , 3	6 , 3
1 , 4	2 , 4	3 , 4	4 , 4	5 , 4	6 , 4
1 , 5	2 , 5	3 , 5	4 , 5	5 , 5	6 , 5
1 , 6	2 , 6	3 , 6	4 , 6	5 , 6	6 , 6

Of 36 possible occurrences, 11 are favorable, so the odds are 11 out of 36. This is equivalent to the probability of rolling one six plus the probability of rolling another six minus the probability of rolling two sixes: $\frac{1}{6} + \frac{1}{6} - \frac{1}{36} = \frac{11}{36}$.

Another way to calculate this is to figure the probability of not getting a 6 on either roll, and subtract that from 1: $1 - (\frac{5}{6} \times \frac{5}{6}) = \frac{11}{36}$.

- What are the odds of tossing at least one head in two tosses of a coin? To answer the question, we can enumerate all possible occurrences as follows:

  H H or H T or T H or T T

  Of 4 possible occurrences, 3 are favorable, so the odds are 3 out of 4. This is equivalent to the probability of tossing one head plus the probability of tossing another head minus the probability of tossing two heads: $\frac{1}{2} + \frac{1}{2} - \frac{1}{4} = \frac{3}{4}$.

  An alternative way to calculate this probability is to calculate the probability of not getting a head and subtracting it from 1: $1\text{-}P(\sim H \bullet \sim H) = 1\text{-}(\frac{1}{2} * \frac{1}{2}) = 1 - \frac{1}{4} = \frac{3}{4}$

Exercises:

1. What is the probability of getting three aces in three successive draws from a deck of cards (a) if each card drawn is replaced before the next drawing is made? (b) if the cards drawn are not replaced?

2. What is the probability of getting no tails in three tosses of a coin?

3. A standard pack of dominoes contains 28 dominoes, 7 of which are doubles: double-six, double-five, double-four, double-three, double-two, double-one, and double-blank. If you draw two dominoes, not replacing the first one, what are the odds you have drawn: (a) two doubles? (b) at least one double?

4. If you draw two dominoes, not replacing the first one, what are the odds that: (a) one of them is a double-six? (b) neither of them is a double-six?

5. If you are dealt two cards from a standard deck of 52 cards, what is the probability that? (a) they will both be black aces? (b) neither of them will be black aces?

6. Four men whose houses are built around a square spend an evening celebrating in the center of the square. At the end of the celebration, each staggers off to one of the houses,

no two going to the same house. What is the probability that each one reached his own house?

7. Suppose that you are dealt four cards from a standard deck of 52 cards. What is the probability that you have: (a) four aces? (b) at least one ace? (c) exactly one ace?

8. In the game of Yahtzee, five dice are used and each player has three rolls. All five dice are rolled on the first roll. On the second roll, a player may pick up any of the dice and roll those dice again. Suppose on Joe's first roll he rolls 2-2-4-4-6. Consider these two options: (a) He picks up the 6 and tries to complete his full house by rolling a 2 or a 4; (b) He picks up the two 2s and the 6, rolling all three dice to try to get three of a kind (4-4-4). Which option gives him better odds? By how much?

9. In drawing three cards in succession from a standard deck, what is the probability of getting at least one spade (a) if each card is replaced before making the next drawing, (b) if the cards drawn are not replaced?

10. If three balls are selected at random from an urn containing five red, ten white, and fifteen blue balls, what is the probability that they will all be the same color (a) if each ball is replaced before the next one is withdrawn, (b) if the balls selected are not replaced?

11. A player in a draw-poker game is dealt three jacks and two small odd cards. He discards the latter and draws two cards. What is the probability that he improves his hand on the draw? (One way to improve it is to draw another jack to make four-of-a-kind; the other way to improve it is to draw any pair to make a full house.)

Deductive Inference

Statements and Statement Forms

In symbolizing statements, we normally use capital letters to represent individual statements. Thus, the statement "Atlanta is the capital of Georgia and Birmingham is the capital of Alabama" would be symbolized as A•B. This statement is clearly a conjunction. If we wish to symbolize the *form* of the statement, without regard to its content, we could use p•q, which is said to be a *statement form*. By convention, capital letters are used to represent statements, and lower-case letters from the middle of the alphabet are used to represent statement forms.

Any statement that is a conjunction is said to be a *substitution instance* of p•q. This is true not only of simple conjunctions such as A•B, but also more elaborate ones such as (A→B)•~(B→C).

All statements, of whatever complexity, are substitution instances of the *generic form p*, as well as of more specific forms. Thus, one sentence may be a substitution instance of many forms.

Exercises On the next page is a list of statement forms. Each of the numbered statements is a substitution instance of one or more of the forms. For each statement, identify all the forms of which it is a substitution instance. The first one is done for you as an example. (Note: some of the forms will be used more than once, some not at all.)

1. 1. C•~D

   This is a substitution instance of a, b, and g.

2. H → (I•J)
3. ~E → (FvG)
4. (K•L) v (M•N)
5. (OvP) → (Q•R)
6. (W → X) → (Y → Z)
7. Z ↔~~Z
8. [(A → B) • A] → B
9. (A → B) • (A•~B)
10. ~(A → B) • (~AvB)
11. (B → A) • (A → B)
12. (A•B) v C
13. (AvC) • (BvC)
14. (A•B) v (A →~B)
15. (A•~B) → (Bv~C)
16. (Dv~H) → [R• (SvT)]
17. (LvM) → (D•E)

(a) p	(d) $p \rightarrow q$	(g) p•~q	(j) $(p \rightarrow q) \rightarrow r$
(b) p•q	(e) $p \leftrightarrow q$	(h) $\sim p \rightarrow q$	(k) $p \rightarrow (q v r)$
(c) p v q	(f) ~p	(i) $p \rightarrow \sim q$	(l) pv(q•r)

Tautologies, Contradictions, and Contingent Statements

If a statement form has only true substitution instances, it is said to be a tautology. For example, it is always true that, whatever X is, either there are X's or there aren't. We can symbolize this statement form as p v ~p and use a truth table to show that it is tautologous.

p	~p	pv ~p
T	F	T
F	T	T

Some tautologies are not as obvious, e.g., $p \rightarrow (q \rightarrow p)$.

p	q	$q \rightarrow p$	$p \rightarrow (q \rightarrow p)$
T	T	T	T
T	F	T	T
F	T	F	T
F	F	T	T

If a statement form has only false substitution instances, it is said to be a contradiction. For example, it is a contradiction to say that something is and is not, at the same time and in the same sense: p•~p.

p	~p	p•~p
T	F	F
F	T	F

Some contradictions are not as obvious, e.g., $(p \rightarrow q) \bullet (p \bullet \sim q)$.

p	q	~q	p→q	p•~q	(p→q) • (p•~q)
T	T	F	T	F	F
T	F	T	F	T	F
F	T	F	T	F	F
F	F	T	T	F	F

Statement forms that have both true and false substitution instances are said to be contingent. Most statement forms fall into this category.

Exercises: Symbolize each of the following statements. Use truth tables to determine whether they are contingent, tautologous, or self-contradictory.

1. If you don't go, then you don't go.
2. Whenever it rains, it either rains or snows.
3. If Mary goes to the party, then her lamb will either go or not go.
4. If it's both raining and not raining, then it is either snowing or not snowing.
5. If I followed your instructions, then I'd both have to turn left at the next light and not turn left at the next light.
6. Whenever I get hungry, I'm not hungry.
7. To be eligible to vote in the United States, you must be at least eighteen years of age.
8. If Boudreaux likes Wagner, then Boudreaux likes Wagner.
9. If Thibodeaux likes Wagner, then he likes opera.
10. Thibodeaux likes both Verdi and Wagner.
11. $[p \rightarrow (p \rightarrow q)] \rightarrow q$
12. $p \rightarrow [(p \rightarrow q) \rightarrow q]$

13. $p \rightarrow [\sim p \rightarrow (q \vee \sim q)]$

14. $(p \bullet q) \bullet (p \rightarrow \sim q)$

15. $p \rightarrow [p \rightarrow (q \bullet \sim q)]$

16. $(p \rightarrow q) \leftrightarrow (\sim q \rightarrow \sim p)$

17. $(p \rightarrow q) \leftrightarrow (\sim p \rightarrow \sim q)$

Logical Equivalence and Logical Implication

The ↔ operator is usually called an equivalence operator. However, what we normally mean by equivalence is that two statements have the same meaning. If the two statements do indeed have the same meaning, then the statement of their equivalence will be a tautology. For example, p•q is obviously equivalent to q•p.

p	q	p•q	q•p	(p•q) ↔ (q•p)
T	T	T	T	T
T	F	F	F	T
F	T	F	F	T
F	F	F	F	T

Two statements that have the same meaning in this sense are said to be logically equivalent. This is analogous to the numeric = operator. We can say, for example, 3+5=7, but such a statement is false. If the statement of equality is true, however, the two expressions are truly equal. Similarly, if the statement of an equivalence is tautological, then the two statements are truly equivalent (i.e., logically equivalent).

Similarly, we use the → operator to symbolize implication. But just because we say that one statement implies another doesn't mean that it really does. If the implication is genuine – if it is a logical implication – then the statement of the implication will be a tautology. For example, the following argument is obviously valid:

p→q
p___
q

When we say that an argument is valid, what we mean is that the premises imply the conclusion. So the argument can be restated as an expression: [(p→q) •p] →q. A truth table will show that the argument is valid:

p	q	p→q	(p→q)•p	[(p→q) • q]→q
T	T	T	T	T
T	F	F	F	T
F	T	T	F	T
F	F	T	F	T

Equivalence Exercises: For each pair of statement forms, use a truth table to determine if form A is equivalent to form B.

A	B
1. p → q	~p →~q
2. p → q	~q →~p
3. (p → q) → r	(q → p) → r
4. p	p • (p v q)
5. p	p v (p • q)
6. p	p • (p → q)
7. p	p • (q → p)
8. p • (q v r)	(p • q) v (p • r)
9. p • (q v r)	(p v q)•(p v r)
10. p → q	(p v q) ↔ q
11. (p • q) → r	p → (q → r)
12. (p → q) • (q → p)	(p • q) v (~p • ~q)

Implication Exercises: For each pair of statement forms, use a truth table to determine if form A implies form B.

A	B
1. (p → q) •~q	~p
2. (p → q) •~p	~q

3.	$(p \rightarrow q) \cdot (q \rightarrow r)$	$p \rightarrow r$
4.	$(p \cdot q) \vee \sim p$	q
5.	$(p \vee q) \cdot \sim p$	q
6.	$p \vee q$	p
7.	$p \cdot q$	p
8.	p	$p \vee q$
9.	$(p \cdot q) \rightarrow r$	$p \rightarrow (q \rightarrow r)$
10.	$(p \cdot q) \rightarrow r$	$q \rightarrow (p \rightarrow r)$
11.	$[(p \cdot q) \rightarrow (r \cdot s)] \cdot (p \vee r)$	$q \vee s$
12.	$[(p \rightarrow q) \cdot (r \rightarrow s)] \cdot (p \vee r)$	$q \vee s$

Assigning Truth-Values

We can use a truth table to show all possible combinations of truth-values when we do not know whether the constituent statements of an expression are true or false. However, if we do know the truth-values of the constituent statements, we do not need to construct a truth table. Instead, we can assign truth-values to the statements and evaluate the expression. Consider, for example, the statement: The Vatican is located in Italy or Sicily. This can be symbolized as I v S. If we constructed a truth table for this expression, it would look like this:

I	S	I v S
T	T	T
T	F	T
F	T	T
F	F	F

However, since we know that I is true and S is false, we can evaluate the expression by assigning truth-values to its constituent statements: T v F, resulting in T for the expression. What we have done, in effect, is construct the second row in the truth table without having to construct the entire table.

Exercises: Determine the truth-value of the following compound statements by assigning truth-values to their constituent statements.

1. London is the capital of England but Stockholm is not the capital of Norway. (L: London is the capital of England. S: Stockholm is the capital of Norway.)
2. If Stockholm is in Sweden, then it can't be the capital of Norway. (S: Stockholm is in Sweden N: Stockholm is the capital of Norway.)
3. Mice are mammals and they don't lay eggs. (M: Mice are mammals. L: Mice lay eggs.)
4. If mice are mammals, then they don't lay eggs. (M: Mice are mammals. L: Mice lay eggs.)
5. If mice don't lay eggs, then they are mammals. (M: Mice are mammals. L: Mice lay eggs.)
6. Lincoln and Washington were both assassinated. (L: Lincoln was assassinated. W: Washington was assassinated.)
7. It's not true that Lincoln and Washington were both assassinated. (L: Lincoln was assassinated. W: Washington was assassinated.)
8. If Lincoln and Washington were both assassinated, then they must both be dead. (L: Lincoln was assassinated. W: Washington was assassinated. L2: Lincoln is dead. W2: Washington is dead.)
9. If Lincoln and Washington are both dead, then they must have both been assassinated. (L: Lincoln was assassinated. W: Washington was assassinated. L2: Lincoln is dead. W2: Washington is dead.)
10. If New Orleans is in Louisiana, and Louisiana is in the United States, then New Orleans must be in the United States. (N: New Orleans is in Louisiana. L: Louisiana is in the United States. U: New Orleans is in the United States.)

11. If New Orleans is in Texas, and Texas is in the United States, then New Orleans must be in the United States. (N2: New Orleans is in Texas. T: Texas is in the United States. U: New Orleans is in the United States.)
12. If New Orleans is in Louisiana but not in Texas, then New Orleans must be in the United States. (N: New Orleans is in Louisiana. N2: New Orleans is in Texas. U: New Orleans is in the United States.)
13. New Orleans is in Louisiana but not in Texas, but New Orleans is in the United States. (N: New Orleans is in Louisiana. N2: New Orleans is in Texas. U: New Orleans is in the United States.)
14. If New Orleans is in Texas, then Bill Clinton is the Queen of Romania. (N2: New Orleans is in Texas. B: Bill Clinton is the Queen of Romania.)

Proving Validity

Truth tables can be used not only to determine the truth-value of an expression, but also to determine if an argument is valid. But in order to do this, we must first make a statement out of the argument. The idea behind this technique is that if an argument is valid, the conjunction of its premises implies the conclusion: $(P_1 \bullet P_2 \bullet \ldots \bullet P_n) \rightarrow C$. We can then construct a truth table for the expression. If the expression is a tautology, the argument must be valid.

Example: If Edwards is honest, then I'm a monkey's uncle. But I'm not a monkey's uncle. Therefore Edwards is not honest.

$E \rightarrow M$ $\quad\quad$ $[(E \rightarrow M) \bullet \sim M] \rightarrow \sim E$

$\underline{\sim M}$

$\sim E$

E	M	~E	~M	E → M	(E → M) •~M	[(E→M) •~M] →~E
T	T	F	F	T	F	T
T	F	F	T	F	F	T
F	T	T	F	T	F	T
F	F	T	T	T	T	T

Another way to determine whether an argument is valid is to leave the premises and conclusion in separate columns. Then we look for the rows in which all the premises are true. If the conclusion is false in any of those rows, the argument is invalid.

Consider the previous argument:

		C	P_2	P_1
E	M	~E	~M	E→M
T	T	F	F	T
T	F	F	T	F
F	T	T	F	T
F	F	T	T	T

Both premises are true only in the last row, and the conclusion is true there also, so the argument is valid.

While this method is theoretically satisfactory for any argument, it becomes extremely cumbersome for longer arguments or arguments with many terms.

Exercises: Use truth tables to prove that the following arguments are invalid.

1. p → q ∴ ~p →~q
2. p → q, ~p ∴ ~q
3. p → q, p → r ∴ qvr
4. (pvq) → (p•q), p•q ∴ ~(pvq)
5. pvq, p ∴ ~q

Proving Invalidity

Now consider a slightly different version of the argument: If Edwards is honest, then I'm a monkey's uncle. But Edwards is not honest. Therefore I'm not a monkey's uncle.

E → M
~E
~M

		P_2	C	P_1
E	M	~E	~M	E→M
T	T	F	F	T
T	F	F	T	F
F	T	T	F	T
F	F	T	T	T

Both premises are true in the third and fourth rows, but the conclusion is false in the third row, so the argument is invalid.

Rather than constructing the entire truth table, we can attempt to construct the row that proves invalidity. We do this by assigning truth-values to the component statements so as to make the premises true and the conclusion false:

E→M, ~E ∴ ~M
F→T ~F ~T
T T F

Take a more complicated example: (H•K) → L ∴ L → (K•H)

Start with the conclusion: how can we make it false? L must be true and (K•H) must be false. Unfortunately, there are three ways to make it false: (T•F), (F•T), and (F•F). So we use trial and error. Let's start with (T•F)

(H•K) → L ∴ L → (K•H)

(F•T) → T ∴ T → (T•F)

F → T ∴ T → F

T ∴ F

We got lucky! Our first assignment of truth-values worked. The argument is invalid because we were able to make a consistent assignment of truth-values such that the premise was true and the conclusion false.

Exercises: This time use the method of assignment to prove that the following arguments are invalid.

1. $p \rightarrow q \;\; \therefore \sim p \rightarrow \sim q$
2. $p \rightarrow q, \sim p \;\; \therefore \sim q$
3. $p \rightarrow q, p \rightarrow r \;\; \therefore qvr$
4. $(pvq) \rightarrow (p \bullet q), p \bullet q \;\; \therefore \sim (pvq)$
5. $pvq, p \;\; \therefore \sim q$

Proving Validity (Re-visited)

The method of assigning truth values can also be used to prove that an argument is valid. The technique is to attempt to assign truth values so that the premises are all true and the conclusion false. If we cannot do so without inconsistent assignment of truth values, the argument must be valid.

Example:

$A \rightarrow B$	$A \rightarrow B$	$B \rightarrow C$	$A \rightarrow C$
$\underline{B \rightarrow C}$	$T \rightarrow T$	$F \rightarrow F$	$T \rightarrow F$
$A \rightarrow C$	T	T	F

The attempt to make the argument invalid required the assignment of inconsistent truth-values to B, therefore the argument is valid.

Review – Methods for Determining Validity

In our study of modern logic, we have used three techniques to determine if an argument is valid:

1. Make the argument into a statement: $(P_1 \bullet P_2 \bullet \ldots \bullet P_n) \rightarrow C$, then use a truth table to evaluate the statement.

- If the statement is a tautology, the argument is valid.
- Otherwise it is invalid.

2. Use a truth table to set up a separate column for each premise and the conclusion (labeled P_1, P_2, … P_n, C). Check the rows in which all the premises are true.
   - If the conclusion is false in any one of these rows, the argument is invalid.
   - Otherwise it is valid.

3. Assign truth-values so as to make all the premises true and the conclusion false.
   - If you can do so without inconsistency, the argument is invalid.
   - If an inconsistent assignment is necessary (when you have tried all options), then the argument is valid. Note that this is in effect a *reductio ad absurdum* argument.

Exercises: Use the method of assignment to determine whether the following arguments are valid or invalid.

1. If Boudreaux goes to the party, then Thibodeaux will not go. Thibodeaux is not going. Therefore, Boudreaux will.
2. Whenever I wash my car it rains. I wash my car. Therefore, it will rain.
3. If women live longer than men, then my neighbor Lauren will live longer than me. Women do live longer than men. Therefore, my neighbor Lauren will live longer than me.
4. The auto dealer says that I can't get both automatic shift and power steering. If I get power steering, then I also must get cruise-control. But I don't want cruise control. So I guess I'll have to get the power steering.
5. I can't afford both power steering and cruise control. I also can't afford both cruise control and power windows. But I really want those power windows. So I guess I can afford neither power steering nor cruise control.

6. If you major in philosophy, you'll have to do a lot of thinking. If you major in engineering, you'll have to do a lot of math. I know that you'd prefer not to spend your time thinking or doing a lot of math. Therefore, I advise you to major in something else.

7. A → B, B →~C ∴ A →~C

8. A → B, C →~B ∴ A →~C

9. A → B, C → D, A v D ∴ B v C

10. (HvI) → [J• (K•L)], H ∴ K•L

11. I v ~J, ~(~K•L), ~(~I•~L) ∴ ~J → K

12. ~(E•F), (~E•~F) → (G•H), H → G ∴ G

13. A•B, (AvC) → D ∴ A•D

14. S → (T → U), V → (W → X), T → (V•W), ~(T•X) ∴ S ↔ U

15. I → J, J → K, L → M, IvL ∴ KvM

16. (Q v ~R) v S, ~Qv (R•~Q) ∴ R → S

17. A ↔ (BvC), B ↔ (CvA), C ↔ (AvB), ~A ∴ BvC

18. T• (UvV), T → [U → (W•X)], (T•V) →~(WvX) ∴ W ↔ X

Formal Proof of Validity

Arguments and Argument Forms

Just as any statement can be regarded as a substitution instance of a statement form, every argument can be regarded as a substitution instance of an *argument form*. Consider the familiar argument:

If Edwards is honest, then I'm a monkey's uncle.
<u>I am not a monkey's uncle.</u>
Edwards is not honest.

If we wanted to capture the form of this argument, we would do so by the use of lower-case letters from the middle of the alphabet, just as we did with statement forms:

$p \rightarrow q$
$\underline{\sim q}$
$\sim p$

Just as names were given to the basic patterns of correct and incorrect reasoning in syllogistic logic, likewise there are names for the basic argument forms in statement (or sentential) logic. Even though there are minor disagreements among logicians about just which argument forms should be regarded as basic, most lists will look something like this:

1. Modus Ponens (MP):
   $p \rightarrow q$
   p
   $\therefore q$

2. Modus Tollens (MT):
   $p \rightarrow q$
   $\sim q$
   $\therefore \sim p$

3. Conjunction (Conj):
   p
   q
   $\therefore p \bullet q$

4. Hypothetical Syllogism (HS):
   $p \rightarrow q$
   $q \rightarrow r$
   $\therefore p \rightarrow r$

5. Disjunctive Syllogism

pvq	pvq
~p	~q
∴ q	∴ p

6. Simplification (Simp):

p•q	p•q
∴ p	∴ q

7. Addition (Add)
   p
   ∴ pvq

8. Constructive Dilemma (CD)
   p → q
   r → s
   pvr
   ∴ qvs

My example argument is obviously a substitution instance of the form known as modus tollens. But just as with statements and statement forms, an argument of considerable complexity can also be a *substitution instance* of a simple form. Consider the following argument:

[A•(B →~C)] → (Cv~D)
<u>~(Cv~D)</u>
~[A•(B→~C)]

This is also an example of modus tollens because it matches the pattern: the expression [A•(B→~C)] can be substituted for p, and (Cv~D) can be substituted for q.

The concept of pattern-matching is fundamental to much of modern logic. It is one reason why computers are so effective as logical tools. Constructing a proof, as we shall see, requires some ingenuity; but recognizing a pattern in a proof that has already been constructed is almost rote work. Well, it's rote work for computers, though it may seem challenging at first to students.

<u>Exercises</u>: Identify each one of the following arguments as a substitution instance of one of the eight basic argument forms (rules of inference).

1. A → C
   C → A
   ∴ A → A

2. A → C
   C → A
   ∴ (A → C) • (C → A)

3. J• (K•L)
   ∴ K•L

4. (HvI) → [J•(K•L)]
   HvI
   ∴ J•(K•L)

5. W•X
   ∴ (W•X) v (~W•~X)

6. (T•U) → (W•X)
   (T•V) → (~W•~X)
   (T•U) v (T•V)
   ∴ (W•X) v (~W•~X)

Formal Proof

In sentential logic, as opposed to syllogistic logic, there is no set number of premises, nor any predetermined order. Also, the method of proof is somewhat different. If the conclusion cannot be inferred directly from the premises (by reference to the basic argument forms), then intermediate statements are generated. These intermediate statements are then used as additional premises, much as the conclusion of a constituent syllogism in a sorites was used as a premise for the next constituent syllogism.

In constructing a formal proof of the validity of an argument, it is common practice to number the premises for the sake of reference – not just the initial premises, but also any intermediate ones. Any numbered line other than the initial premises must be justified. A justification consists of the line numbers for the premises from which this line is inferred, along with the abbreviation for the basic argument form that is used. This process continues until the final line of the proof is identical with the conclusion of the argument.

Now for a bit of clarification. Even though the order of the premises is unimportant, order is important for a justification since we are trying to match the pattern of an argument form. For example, suppose we have the following argument:

1. B → C	
2. A → B	/∴ A → C
3. A → C	1, 2 HS

The justification is incorrect because it doesn't quite match the pattern.

p → q	A → B	B → C
q → r	B → C	A → B
∴p → r	∴ A → C	∴ A → C
pattern	correct match	incorrect match

It's important to understand the difference between an inference and a justification. All the inferences above are correct, because they are all valid arguments. However, the last one is not a correct justification because it doesn't match the pattern.

The justification should be 2,1 HS.

Perhaps a couple of examples of complete proofs will be helpful.

1. Av (B → D)	
2. ~C → (D → E)	
3. A → C	
4. ~C	/ ∴B → E
5. ~A	3,4 MT
6. B→D	1, 5 DS
7. D→E	2, 4 MP
8. B→E	6, 7 HS

1. (A•B) → [A→ (D•E)]	
2. (A•B) •C	/ ∴ D v E
3. A•B	2 Simp
4. A→ (D•E)	1, 3 MP
5. A	3. Simp
6. D•E	4, 5 MP
7. D	6, Simp
8. DvE	7, Add

These examples illustrate several points. One is that the same premise may be used more than once. Another is that there's no one right way to construct a proof. For example, in the first proof, lines 6 and 7 could have been done in reverse order without affecting the outcome. Another point is illustrated in the second proof, and that is that sometimes you will need to use the same argument form repeatedly in the same proof.

Exercises: In the following formal proofs of validity, supply the justification for each line that is not a premise. Note: these proofs use only the eight basic argument forms (rules of inference).

(1)
1. A → B
2. B →~C /∴ A →~C
3. A →~C

(2)
1. (HvI) → [J•(K•L)]
2. H /∴ K•L
3. HvI
4. J• (K•L)
5. K•L

(3)
1. A•B
2. (AvC) → D /∴ A•D
3. A
4. AvC
5. D
6. A•D

(4)
1. I → J
2. J → K
3. L → M
4. IvL /∴ KvM
5. I → K
6. KvM

(5)
1. Fv(GvH)
2. (G → I) • (H→J)
3. (IvJ) → (FvH)
4. ~F /∴ H
5. G → I
6. H → J
7. GvH
8. IvJ
9. FvH
10. H

Rules of Replacement

There are many intuitively valid arguments that cannot be shown to be valid by use of the eight basic argument forms. Consider the following argument:

> You know that if either John or Sue goes to the party, then Ann won't. And I'm sure that Sue or John will go. So Ann won't.

The problem is obvious if we symbolize the argument, because it doesn't match the pattern for modus ponens:

(JvS) →~A	p→q
SvJ	p
~A	q

It doesn't match because SvJ is not identical to JvS. But we all know that the two statements are logically equivalent, just as we all know that 3+4 equals 4+3. But a formal proof of validity cannot contain a justification such as "everybody knows this". Formal proof is based on pattern matching, not on what "everybody knows".

What we have to do is to add a pattern that would justify substituting JvS for SvJ. This pattern is known as commutation: (pvq)↔(qvp). But note the difference between this rule and the earlier ones. The equivalence operator indicates two-way inference. It also permits us to replace a part of a statement with an equivalent expression. Thus, instead of replacing S v J with J v S, we could just as well replace (J v S)→~A with (S v J)→~A. Either replacement is justified by the rule of commutation. Moreover, commutation is valid for conjunctions as well as disjunctions: (p•q)↔(q•p).

When we are doing arithmetic, the order of computation can sometimes affect the result. For example, if we have the expression 3+5×2, we will get 16 if we do the addition first, but 13 if we do the multiplication first. We can use parentheses to indicate the order of computation: (3+5)×2 or 3+(5×2). But sometimes the order of computation is unimportant: (3+5)+2 gives us the same result as 3+(5+2). We can use a similar rule in logic when the order of operation makes no difference. This rule is known as association: [pv(qvr)]↔ [(pvq)vr]. And there is a similar pattern for conjunctions: [p•(q•r)]↔[(p•q)•r].

Over the years logicians have come up with additional rules that are needed in order to be able to demonstrate the validity of arguments. What is generally accepted today is a set of ten rules of replacement to supplement the eight basic argument forms. (You will recognize some of these from the earlier discussion of Boolean algebra.) The basic argument forms are also known as rules of one-way inference, and the rules of replacement are also known as rules of two-way inference.

9. Double Negation (DN) – 2 forms

p	~~p
∴ ~~p	∴ p

10. Exportation (Exp) – 2 forms

(p•q) → r	p → (q → r)
∴ p → (q → r)	∴ (p•q) → r

11. Commutation (Comm) – 4 forms

pvq	qvp
∴ qvp	∴ pvq
p•q	q•p
∴ q•p	∴ p•q

12. Association (Assoc) – 4 forms

p v (q v r)	(p v q)v r
∴ (p v q) v r	∴ p v (q v r)
p•(q•r)	(p•q)•r
∴ (p•q)•r	∴ p•(q•r)

13. Distribution (Dist) – 4 forms

p•(q vr)	(p•q) v (p•r)
∴(p•q) v (p•r)	∴ p•(q v r)
p v (q•r)	(p v q) • (p v r)
∴(p v q)•(p v r)	∴ pv(q•r)

14. Contraposition (Contra) – 2 forms

p → q	~q →~p
∴ ~q → ~p	∴ p → q

15. Implication (Impl) – 2 forms

p → q	~pvq
∴ ~pvq	∴ p → q

16. DeMorgan's Theorem (DeM) –4 forms

~(p•q)	~pv~q
∴ ~pv~q	∴ ~(p•q)
~(pvq)	~p•~q
∴ ~p•~q	∴ ~(pvq)

17. Tautology (Taut) – 4 forms

p	p•p
∴ p•p	∴ p
p	p v p
∴ p v p	∴ p

18. Equivalence (Equiv) – 4 forms

p ↔ q	(p → q) • (q → p)
∴ (p → q)•(q → p)	∴ p↔q
p ↔ q	(p•q) v (~p•~q)
∴ (p•q) v (~p•~q)	∴ p ↔ q

Exercises: Identify the rules of replacement that are used in the following arguments.

1. (I v ~~J) •K
   ∴ (IvJ) •K
2. ~L →~(K•J)
   ∴ (K•J) → L
3. K → (I →~M)
   ∴ (K•I) →~M)
4. K• (IvJ)
   ∴ (K•I) v (K•J)
5. ~Mv~~L
   ∴ ~(M•~L)

6. (IvJ) •K
   ∴ K• (IvJ)
7. (A → C) • (C → A)
   ∴ A ↔ C
8. A ↔ C
   ∴ (A•C) v (~A•~C)
9. ~(A•C)
   ∴ ~A v ~C

Constructing Proofs

As I indicated earlier, constructing a formal proof is not a rote task. It requires some creativity and ingenuity. Like other skills, it can be acquired through practice. The first step is to look at formal proofs that others have created, and try to learn from their examples (much as one can begin to learn how to write well by reading good writing).

<u>Exercises</u>: In the following formal proofs of validity, supply the justification for each line that is not a premise. Note: these proofs use both the rules of inference and the rules of replacement.

(1) 1. A → B
2. C →~B   /∴ A → ~C
3. ~~B →~C
4. B →~C
5. A →~C

(2) 1. T• (UvV)
2. T → [U → (W•X)]
3. (T•V) →~(WvX) /∴ W↔X
4. (T•U) → (W•X)
5. (T•V) → (~W•~X)
6. (T•U) v (T•V)
7. (W•X) v (~W•~X)
8. W ↔ X

(4) 1. F
2. G → H
3. I → J
4. (HvJ) →~F/∴ ~G•~I
5. ~~F
6. ~(HvJ)
7. ~H•~J
8. ~H
9. ~G
10. ~J
11. ~I
12. ~G•~I

(3) 1. (Qv~R)vS
2. ~Qv(R•~Q) /∴ R → S
3. (~QvR) • (~Qv~Q)
4. ~Qv~Q
5. ~Q
6. Qv(~RvS)
7. ~RvS
8. R → S

(5) 1. A → B
2. B → C
3. C → A
4. A →~C/∴~A•~C
5. A → C
6. (A → C) • (C → A)
7. A ↔ C
8. (A•C) v (~A•~C)
9. ~Av~C
10. ~(A•C)
11. ~A•~C

It's important to note that a failure to construct a formal proof is no evidence of invalidity; it may instead indicate a lack of ingenuity. To show that an argument is invalid, it would be necessary to use one of the techniques discussed earlier, such as the method of assignment. For our present purposes, we will assume that all of the arguments are valid.

There is no mechanical technique for constructing a proof, but there are some techniques that are often useful. One that I have found to be most helpful is to work back from the conclusion, trying to find what would be necessary in order to produce it. Let's consider an example:

1. A → B
2. ~(BvC)
3. D → (CvA)
4. E → (DvF)
5. ~F /∴~E

There's only one place the conclusion can come from, and that is statement 4. So we need to ask ourselves, how can we get ~E out of statement 4? The obvious answer is by the use of modus tollens:

E → (DvF)
~(DvF)
~E

So we can get ~E if we can get ~(DvF). There doesn't seem to be any place to get ~(DvF), so we need to think of its DeMorgan equivalent: ~D•~F. We have ~F in statement 5, but where can we get ~D? We need to think of modus tollens again, this time using statement 3:

D → (CvA)
~(CvA)
~D

So how can we get ~(CvA)? Once again we use DeMorgan to get ~C•~A. We can get ~C from statement 2 by DeMorgan and simplification. Getting ~A is a bit more involved. The previous use of DeMorgan on statement 2 yielded ~B•~C. We can get ~B by simplification, then use it with statement 1 to get ~A by modus tollens.

Now we are ready to put it all together:

1. A → B	
2. ~(BvC)	
3. D → (CvA)	
4. E → (DvF)	
5. ~F	/∴ ~E
6. ~B•~C	2 DeM
7. ~B	6 Simp
8. ~A	1,7 MT
9. ~C	6 Simp
10. ~C •~A	9,8 Conj
11. ~(CvA)	10 DeM
12. ~D	3,11 MT
13. ~D •~F	12,5 Conj
14. ~(DvF)	13 DeM
15. ~E	4,14 MT

Exercises, part 1: Proofs for the following arguments can be constructed using only one-way rules of inference.

1. 1. B → C
   2. B /∴ B•C

2. 1. B → C
   2. D → E
   3. A → B
   4. C → D /∴ A→E

3. 1. ~C
   2. A → C

3. AvB /∴ BvD

4. 1. A •(B•C) /∴ A•C

5. 1. A
 2. C /∴ (AvB) • (CvD)

6. 1. (H•K) → L
 2. ~R•K
 3. K→(HvR) /∴ L

7. 1. A → B
 2. ~B
 3. (~Av~C) → (D•E) /∴ EvF

Exercises, part 2: In constructing proofs for the following arguments, you will need to use the rules of replacement as well as the rules of one-way inference.

1. 1. ~Av (B•C) /∴ A→C

2. 1. (MvN) → (O•P)
 2. ~O /∴ ~M

3. 1. (NvF) → (C•D)
 2. D → V
 3. V → I
 4. I → A
 5. A → ~C /∴ ~F

4. 1. A → B
 2. C →~B /∴ ~(A•C)

5. 1. A →~B
 2. AvC
 3. B•D /∴ C

6. 1. X ↔ Y
 2. ~X /∴ Y → Z

7. 1. [Av (B•C)] → (D•E)
 2. ~E /∴ ~A

8. 1. ~Kv(~Lv~M) /∴ (K•L) →~M

9. 1. $(O \bullet T) \rightarrow (S \rightarrow M)$
   2. $R \rightarrow \sim M$
   3. $T \bullet R$
   4. $O \bullet S$ $\quad /\therefore\ V$

10. 1. $L \rightarrow M$
    2. $K \rightarrow P$
    3. $(M v P) \rightarrow Z$
    4. $\sim Z$ $\quad /\therefore\ \sim(L v K)$

11. 1. $(D v \sim H) \rightarrow [R v (S \bullet \sim T)]$
    2. L
    3. $(L v M) \rightarrow (D \bullet E)$ $\quad /\therefore\ T \rightarrow R$

12. 1. $N \rightarrow O$
    2. $(N \bullet O) \rightarrow P$
    3. $\sim(N \bullet P)$ $\quad /\therefore\ \sim N$

Inductive Logic

Inductive Inference

Logic is a very large topic. Like other very large topics, it is frequently divided into sub-topics. One of the customary divisions is deductive vs. inductive inference.

There are various ways to characterize the difference between these two categories of inference. One way to differentiate between the two is on the basis of whether the inference provides any new information. Deductive inference does not. It "draws out" or makes explicit the information that was already there, at least implicitly. Consider, for example, the syllogistic argument:

> Dogs are mammals.
> Beagles are dogs.
> Beagles are mammals.

The conclusion of this argument does not add anything to the storehouse of information that was already available. The process of inference simply makes explicit what was already contained implicitly in the premises.

Now consider a typical inductive argument:

> I like all of Verdi's operas that I have seen, so I'll probably like *Falstaff*, which I haven't seen.

The conclusion of this argument, "I'll probably like *Falstaff*", provides new information. I know I like *Aida* and *Rigoletto* – I've already seen them – but I've never seen *Falstaff*, so I don't know if I will like it. But the argument says I probably will.

The way this argument is worded gives us another clue to the difference between deductive and inductive inference, and that is the degree of confidence that is warranted. With deductive inference, if the argument is valid, the conclusion follows from the premises *with certainty*. It does so because the conclusion is already contained implicitly in the premises, and the inference is a process of drawing out what is already there. By contrast, inductive inference is always more or less probable, never certain. The more of Verdi's operas I have seen and liked, the more likely

I will also like *Falstaff.* But I might not. I have heard that *Falstaff* is a comedy, whereas *Aida* and *Rigoletto* and the others that I have seen are tragedies. I can't imagine Verdi doing comedy.

Argument by Analogy

Our earlier discussion of analogy focused on the inappropriate use of analogical argument, such as the fallacy of questionable analogy. But not all analogies are questionable. In fact, analogical argument can be a very powerful tool of inductive inference, as long as we recognize its limitation. My example above about Verdi's operas is a sort of analogical argument. I am comparing *Falstaff,* which I haven't seen, with *Aida* and *Rigoletto,* which I have seen. I am saying that *Falstaff* is in some important ways analogous to *Aida* and *Rigoletto* – that is, composed by the same person – and it is likely that I will enjoy it, as I have enjoyed them. But I also acknowledged that there is at least one very important difference between *Falstaff* and the other operas – it is a comedy, whereas they were tragedies – so that might lessen the probability that I will like it.

Is there any way that we can formalize these insights, so that we will have a set of rules for evaluating analogical arguments? The answer is a guarded "Yes". In working with inductive inference, there is no way we can come up with a definitive set of criteria comparable to the rules for categorical syllogisms. But there are guidelines that are generally useful, as long as we realize that they are not exact.

Let's go back to my opera analogy. There are several ways that my argument – and analogical arguments in general – can be strengthened:

1. Other things being equal, the <u>more instances</u> on which a comparison is based, the greater the strength of the comparison. Consider my argument about probably liking *Falstaff.* This argument is stronger if I have seen and liked ten of Verdi's operas than if I have only seen and liked two of them.

2. The more points of similarity there are in the comparison, the stronger the argument. For example, if all of the Verdi operas I have seen were performed in Italian, and *Falstaff* will also be performed in Italian, that strengthens the probability that I will like it.

3. The more different kinds of experiences that have produced the same result, the stronger my argument. For example, if I have seen Verdi's operas in live performance and on video, in Italian and in English, etc., and have liked them all, the more probable it is that I will like *Falstaff.*

4. The less I claim for my conclusion, the stronger the argument. For example, it is more probable that I will simply like *Falstaff* than that I will think it is the best opera I have ever seen.

Some considerations will weaken an analogical argument. These are called disanalogies. The consideration I mentioned earlier – that *Falstaff* is a comedy, unlike the other operas by Verdi that I have seen – is a disanlogy. There are still other considerations that neither strengthen nor weaken the argument. Suppose someone says to me: "I'm sure you will enjoy seeing *Falstaff* tonight. After all, it's your birthday." Such a consideration is probably irrelevant. I'm not likely to enjoy the opera any more or less because I see it on my birthday.

Exercises: For each of the following arguments by analogy, several additional factors are suggested. For each of the additional factors, decide whether its addition would make the argument more probable, less probable, or have no effect on the argument.

1. For the past four years, Joe has purchased several shares of Pear Computer stock, and each year the stock has appreciated in value by at least 5%. The dividends paid by the stock have varied, but the average has been 6½%. This year Joe decides to buy about the same number of shares as he did last year, thinking he will continue to enjoy good returns on his investment

   a. Suppose that he had bought stock for the past ten years instead of four.

b. Suppose that the stock had appreciated at a rate of at least 10% instead of 5.

c. Suppose that he had read that one of the chief competitors of Pear had gone out of business.

d. Suppose that he had read that Pear's new line of computers will be called the Bartlett.

e. Suppose that he decides to buy stock in a start-up computer company instead of Pear.

2. Although Suzy was bored by most of the operas she saw in the past, she agreed to go with Joe to the opera tonight, expecting to be bored again.

   a. Suppose that the operas she saw in the past were all tragedies, but the opera they plan to see tonight is the famous comedy *The Barber of Seville*.

   b. Suppose that Suzy had been bored by all, not just most, of the operas she had seen in the past.

   c. Suppose that all of the operas that Suzy had seen in the past were sung in Italian, and Joe told her that tonight's opera would also be sung in Italian.

   d. Suppose that Suzy was generally bored by opera regardless of who she went with.

3. Boudreaux loves to go fishing, even if he doesn't catch many fish. When his friend Thibodeaux invites him to go to Holly Beach to go fishing tomorrow, he eagerly accepts, expecting to "pass a good time," as the Cajuns say.

   a. Suppose that Thibodeaux tells him that when he went the day before, he came home with his ice chest full of fish.

   b. Suppose that Thibodeaux tells him that when he went the day before, he didn't catch a thing.

   c. Suppose that Thibodeaux tells him that he will have an ice chest full of beer tomorrow, and Boudreaux loves to drink beer while he fishes.

   d. Suppose that tomorrow is the 4th of July.

Refutation by Logical Analogy

Analogy is a more powerful tool for disproof than proof. Sometimes we can use a simple analogy to absolutely discredit an argument. It works like this: we attempt to show that the position that someone is advocating is like a position that runs counter to common sense or that is generally not accepted. Consider the following example:

> The states of New Jersey and New York have disputed ownership of Ellis Island. This tiny island is located at the mouth of the Hudson River near the Jersey coast. During the 19th century, many thousands of immigrants passed through Ellis Island on their way to the U.S. Some people have argued in favor of New York ownership of the island, since most of those immigrants were bound for New York, not New Jersey. But critics have pointed out that Cincinnati International Airport is located in Covington, Kentucky. Since most people who get off at the airport are bound for Ohio, not Kentucky, the same logic would suggest that Covington should belong to Ohio.

Exercises: Each of the following arguments is intended to be a refutation by logical analogy. In each case, identify the argument that is being refuted. Also identify the refuting analogy, then determine whether the refuting analogy is sufficiently similar to the argument being refuted to be an effective refutation.

1. Some lawyers and judges have argued against television coverage of trials. They say that cameras in the courtroom make the system look bad. But they are confusing the messenger with the message. That is like saying that the press coverage of the Vietnam war is the reason that the U.S. lost the war. The truth is, if press coverage of something makes it look bad, that is all the more reason for having press coverage.

2. Some people argue that the U.S. is a Christian nation because the framers of the Constitution and early law-makers in the United States were Christians. This is like saying that because ancient Romans believed in a pantheon of

gods, Europeans should today bow at the feet of statues of Jupiter and Juno.

3. The argument against new highways is given forceful statement by three distinguished urban planners who argue as follows: "New highways bring induced traffic. So building more highways will only cause more traffic congestion, not less." This is nonsense. Long lines at a grocery store would not prompt anyone to say, "Well, we can't build any more grocery stores. That would only bring out more customers." Building more highways wouldn't lure cars. The cars come anyway.

4. America's supply of timber has been increasing for decades, and the nation's forests have three times more wood today than in 1920. "We're not running out of wood, so why do we worry so much about recycling paper?" asks Jerry Taylor, the director of natural research studies at the Cato Institute. "Paper is an agricultural product, made from trees grown specifically for paper production. Acting to conserve trees by recycling paper is like acting to conserve cornstalks by cutting back on corn consumption."

Explanation

At the level of everyday discourse, explanations are usually given in answer to "Why" questions. When I ask a guy, "Why did you do that?" I could be asking one of two things:

- <u>Motivation</u>: a psychological question of what he was trying to accomplish.
- <u>Justification</u>: an ethical or legal question of why he considered his behavior appropriate.

We must recognize, however, that it isn't always possible to distinguish between the two. When little Tommy says, "I hit Joe because he hit me first", he may mean that Joe made him angry (motivation) or that Joe deserved payment in kind (justification).

When we are dealing with physical phenomena, explanations are simpler. When a little girl asks, "If the earth is spinning so fast, why don't we fall off?", she wants to know what holds us on. The question is about causes, as are most of our questions of physical explanation.

Causation

Our common-sense notion of causation is not very precise. When we say that A causes B, we mean that if A is present, B will surely follow. But sometimes it doesn't. "Arsenic is poisonous – it will kill you." This is a good rule in general, but there are documented cases of people who have developed a tolerance for arsenic because it occurs naturally in the soil where they live. "Heavy people fall harder than light ones." Well, most of the time, but in a gravity-free environment, they don't. Our common-sense statements about cause and effect assume a certain background without which the cause will not have its normal operation.

We can also distinguish between proximate and remote causes. Ordinarily, we explain things by proximate causes. "Henry was killed because the driver did not see him walking along the roadside" is preferable to a lengthy explanation of the development of the internal combustion engine, its subsequent use in automobiles, etc. But sometimes we use remote explanations. In some sense, everyone dies of cardiac arrest, but we don't use that explanation for the victim of an automobile accident. Also, psychologists and sociologists sometimes resort to remote explanations of deviant behavior: "Joe killed his whole family because he had a psychotic condition caused by years of abuse."

Mill's Methods

In the 19th century, the logician John Stuart Mill developed several techniques for justifying a statement that A causes B.

The Method of Agreement: If two or more instances of the phenomenon under investigation have only one circumstance in common, then this common circumstance is the cause (or effect) of the phenomenon under investigation.

> Example: Suppose several friends have lunch together at a local restaurant. Within an hour, three of them get sick at the stomach, but the others feel fine. Using the method of

agreement, we attempt to determine whether the three who got sick ate the same food, and the others did not. Upon investigation, we find that the three who got sick ate a very spicy *sauce piquante*, which the others did not have. We conclude that their stomach ailments were probably caused by the *sauce piquante*.

The Method of Difference: Suppose you are trying to determine the cause of a given phenomenon. If there are two sets of circumstances that are similar in all respects except one, and in one the phenomenon under investigation occurs and in the other it does not occur, the circumstance in which alone the two instances differ, is the effect, or the cause, or an indispensable part of the cause, of the phenomenon under investigation.

Suppose, in the preceding example, we discover that one person – Joe – who ate the *sauce piquante* did not get sick. We might be inclined at first to think it was not the *sauce piquante* that made the others sick. But before jumping to that conclusion, we inquire whether there is anything different about Joe that would explain why he didn't get sick. Upon investigation, we find that Joe has a reputation for a cast-iron stomach: he can eat almost anything and not get sick. So we have used both the method of agreement and the method of difference to explain why most of the people who ate the *sauce piquante* got sick, but one did not.

The Joint Method of Agreement and Difference: This method consists in the use of both the Method of Agreement and the Method of Difference in the same investigation.

Our first example was really the joint method, because we found that those who got sick consumed the same food (agreement), and those who did not get sick consumed different food (difference). It seemed at first that we had found an exception with Joe, until we found another explanation for the different result in his case.

The Method of Concomitant Variation: A phenomenon that varies in any manner (whether directly or inversely) whenever another phenomenon varies in some particular manner, is either a cause or an effect of that phenomenon (or is somehow causally connected with it).

> In some cases of cause and effect, it is not a question of all or nothing, but of more or less. Let's go back to the *sauce piquante*. Suppose Sue is a picky eater – she ate very little *sauce piquante* and only got slightly sick. By contrast, Harold went back to the buffet line for a second helping, and he got the sickest of all.

We need to recognize that the methods, while helpful, are by no means fool-proof.

- All of the methods assume that the person using the methods has an independent sense of causation. For example, they do not show which of two correlated phenomena is cause and which is effect. Perhaps lung cancer causes smoking.
- The methods will not work unless the cause is among the phenomena under investigation. For example, if I am trying to determine the cause of recurring indigestion, I will probably focus on food. But maybe my indigestion is caused by stress or some other emotional factor. If I only consider food, I will not find the cause.
- If there is more than one common ingredient in all the phenomena under investigation, I may mistakenly choose the wrong one. Recall the case of the scientific drinker. He attempted to determine the cause of his frequent inebriation by carefully noting what he drank. He looked back over his list at the end of the week. Monday night: vodka and orange juice. Tuesday night: gin and orange juice. Wednesday night: bourbon and orange juice. Thursday night: Scotch and orange juice. Friday night: rum and orange juice. By Saturday he was convinced that his inebriation was caused by orange juice.
- Sometimes two things that are statistically correlated are not causally connected to each other, but to a third thing. For example, my health has declined in proportion to my decrease in smoking. Should I therefore conclude that the decrease in smoking has caused the decline in health? I

don't think so. I think both are age-related. I have experienced a natural decline in health as I have grown older, and I have also reduced my consumption of many things, tobacco among them, during the same time period.

Exercises: Analyze each of the following arguments, and indicate which of Mill's methods are being used in each of them.

1. Little Suzy discovers some old paint in the garage. Her favorite color is pink, so she decides to mix some red and white paint. She notices that the more white paint she adds, the lighter the shade of pink. Which method is Suzy using?

2. At his wife's request, Jack replaces a bulb in her favorite lamp, but the lamp still doesn't work. He tries another bulb, with the same result. Jack concludes that the lamp is no good. What else could he do to substantiate his conclusion?

3. I used to suffer from chronic indigestion, but now I rarely have indigestion problems. My theory is that it's because I now drink a glass or two of wine with each meal (except breakfast). What could I do to test my theory?

4. According to many studies, Louisiana has low scores in education, health, and longevity. But other studies show that Louisiana scores high in satisfaction surveys (in the sense that people who live here would rather stay here than move somewhere else). Are the happy people ignorant of the bad things, or do they just not care? How could you find out?

5. One of my favorite radio shows is *Car Talk*. The hosts invite people to call in with car problems which they attempt to diagnose. The questions they ask their callers are of two kinds: (1) Under what conditions does the problem occur? (2) Under what conditions does the problem not occur? Which method (or methods) are they using?

6. A child playing with the volume control on a radio notices that as she turns it one way, it gets quieter, and the other way it gets louder. Which method is the child using?

7. The medicine I buy for my dog has dosage based on weight. I can't weigh the dog directly, so I step on the scales without

the dog, then again with the dog. Which method am I using?

8. It has been claimed that older people who live alone have better health if they have a pet. How could we test this claim? If the claim is verified, how could we determine what kind of pet is most effective (say a cat compared to a goldfish)?

9. The definition of global warming is non-controversial: it is defined as a worldwide increase in average temperature. What is controversial is how much of the change is man-made. What sort of evidence is relevant to resolving this question?

Scientific Method

Inductive Generalization

We have names for people and pets, and sometimes even for inanimate objects such as boats. I remember many years ago listening to a friend talking about Mr. Tuttle as if he were a member of the family or a close family friend. It was only later that I discovered that "Mr. Tuttle" was a name he had given to an heirloom automobile. However, we do not normally name all the individual things of our experience – every cockroach or crawfish, every corn stalk or blade of grass. If we attempted to do so, we would experience what computer technicians call data overload. Instead, we group individuals into categories, and normally use the group name, such as "grass". If we wish to be more specific, we may talk about types of grass, such as Bermuda or St. Augustine.

Grouping similar things is not done only for the purpose of naming things. We also use grouping as the basis for a type of inductive inference known as generalization. If we recognize something as a member of a group, we will assume that it has the same properties as the other members of the group. However, we are not always right. We only need recall the discovery of a strange little animal known as the duck-billed platypus. We normally classify an animal with a bill as a bird, but the platypus is a mammal. Such discoveries serve to remind us that inductive inference is never certain, only more or less probable.

There are two generally-recognized types of generalization, categorical and statistical:

- Categorical generalization: We conclude that, because all observed instances of X have property Y, therefore it is probable that all Xs have property Y. For example, all the crows that I have seen are black, so I infer that all crows are black.
- Statistical generalization: We conclude that, since property Y has been observed to occur in a certain percentage of

> X's, therefore it is likely that property Y will occur in the same percentage rate in all instances of X. This sort of generalization is the basis of statistical surveys. When we read that, for example, "20% of Americans have no idea where Yellowstone National Park is located", we should not suppose that the people who conducted the survey talked to all or even most Americans. Such statements are normally based on a sample population that is supposed to be representative of the population as a whole. If the sample is not representative, then of course the statistic is not valid. We need only recall the infamous telephone poll of 1948, on the basis of which Dewey was predicted to win the Presidency in a landslide. Does anyone remember President Dewey?

Scientific Explanation

Scientists do a lot of grouping and naming, as anyone who has struggled through an introductory biology class will no doubt recall. They also engage in generalization as a way of characterizing the general properties of the groups they have named. But they do a great deal more than that. They also want to provide broad explanations that will find a common cause for many phenomena that at first sight do not appear to have much in common. These broad explanations are known as general theories. Once they have become widely accepted by scientists, they are known as laws. A familiar example is the law of gravity. It finds a common explanation for such apparently unrelated things as apples falling from trees, the rising and falling of the tides, and the movement of the planets around the sun.

Some logicians have claimed that general theories are inductive generalizations, but I don't think that is the case. For one thing, mathematics plays too large a role in the theories of physics and astronomy for these theories to be considered inductive generalizations. Consider the mathematical formulation of the theory of gravity: Every object attracts every other object, directly according to the product of their masses, and inversely according to the square of the distance that separates them. Does anyone really believe that Isaac Newton came up with this formulation by generalizing from his experience of apples and tides and such?

In fact, great scientists are probably as creative as great artists or musicians. Their hypotheses are often the work of inspired genius. But at first they may appear as the ravings of a madman (like Freud's theory of the subconscious). It is important to realize that much science flies in the face of common sense. The central concept of evolution seems to contradict the regularity of nature. Any farm boy can tell you that when two horses mate, they make another horse. How, then, could the species ever change? Or take the physicist's claim that some mass is so dense that a handful would have the weight of the entire earth. Spiritualism makes more sense.

Although modern science is called empirical science, it is important to realize in what sense it is empirical. Few of the important claims of science are directly observable. Many are in principle unobservable. Observation and experimentation are relevant to scientific explanation, but the relationship is not simple. More on this later.

One way to distinguish scientific explanation from ordinary explanation is to focus on what is being explained. Ordinary explanation focuses on irregularities: why John did not show up for work today, not why he came to work all the other days since he was hired. By contrast, science attempts to explain the regularities in nature. It looks for general laws. Moreover, scientists want to find the most general laws possible – what are called unifying theories.

Since scientific explanation is in many ways counter-intuitive, we need some criteria by which we can evaluate any proposed explanation:

- Relevance: This is fundamental. If the explanation is not relevant to what was supposed to be explained, we need go no further.
- Testability: General theories are not directly testable, especially when they postulate unobservable entities such as quarks and leptons. What, then, is the difference between a

scientific explanation and a spiritualistic one? The scientific explanation has testable consequences. It is in principle falsifiable by observation and experiment.

- Compatibility with other scientific laws: This is not an absolute principle, because so-called scientific laws do change over time, but there is a great reluctance to change well-established theories. That is why the change from the astronomy of Ptolemy to that of Copernicus was called a revolution.
- Predictive or explanatory power: Other things being equal, a hypothesis that permits us to predict more is preferred to one that permits less.
- Simplicity: This is the principle of logical economy, otherwise known as Occam's Razor. This is one of the reasons for preferring Copernican over Ptolemaic cosmology. The latter required the assumption of cycles and epicycles for the planets, and Copernicus got rid of the epicycles. I think one way to view the work of Darwin is that it was an attempt to show that the assumption of intelligent design is not necessary to explain the order in nature. For centuries intelligent design was the assumption of laymen and scientists alike. Order in everyday life is associated with intelligent design, so the obvious explanation of order in the cosmos is intelligent design. In fact, the driving force behind the work of many of the great scientists of the past was the belief that God had created the world on a great master plan, and if we keep looking we can find it. But Darwin in effect invoked the principle of logical economy by saying that nature is self-explanatory: evolution takes the place of the master plan.

The Hypothetico-Deductive Technique

Although there is no way to mechanize the creative aspect of scientific theories, we can state the general steps involved in the formulation and testing of a theory. I will use the theory of evolution as an example.

- First, a tentative hypothesis is formulated. How and why it is formulated is irrelevant. It may have been a stroke of genius. Or, like the theory of evolution, it may have been around for centuries before someone figured out how to turn it from fantasy into science.
- Then additional evidence is collected. Darwin did this in his famous voyage of discovery.
- Next, the hypothesis is revised as necessary. Darwin did not understand the mechanism of evolution – how it actually worked – until he discovered the various species of finches on the Galapagos Islands.
- Then comes deducing further consequences. This takes the form of a prediction: if this hypothesis is correct, we should be able to observe so-and-so. Darwin said that if humans descended from apes, as his theory suggested, then we should find fossil evidence of intermediate species.
- Finally, test the consequences: The predictions can be confirmed or disconfirmed by observation and experiment. And in fact fossil evidence continues to be discovered that appears to fill in the gap between ape and man although not in the way that Darwin envisioned it.

Confirmation

Hypotheses are more or less probable, never certain. To confirm an hypothesis is to accumulate evidence for it, that is, to find out how probable it is. All hypotheses are formulated at a specific time in the history of the world. They therefore assume as a background what is known about the world at that time. If we call this background knowledge the antecedent conditions (A), we can symbolize the hypothetico-deductive method as follows $(A \bullet H) \rightarrow O$, which by exportation is equivalent to $A \rightarrow (H \rightarrow O)$. In other words, given the antecedent conditions, if our hypothesis is true, then we should expect certain observations. The pattern of confirmation is:

1. $(A \bullet H) \rightarrow O$
2. $O \quad\quad /\therefore\ H$

Note that as a deductive argument this would be invalid; it is the fallacy of Affirming the Consequent. But as one piece of evidence for a probability claim, it is valid.

The pattern of disconfirmation is similar:

1. (A•H) → O
2. ~O /∴ ~H

Note that this argument is deductively valid, if we assume that the antecedent conditions are true:

1. (A•H) → O	
2. A	
3. ~O	/∴ ~H
4. A → (H→O)	1 Exp
5. H → O	4,2 MP
6. ~H	5,3 MT

However, part of what we call our background knowledge consists of other hypotheses that have been accepted in the past. Therefore a more accurate symbolism is $(H_1 \bullet H_2 \bullet H_3) \rightarrow O$, where H1 is the hypothesis under question, and H_2, H_3, etc. represent generally accepted hypotheses. But if we now get ~O, we should infer $\sim(H_1 \bullet H_2 \bullet H_3)$, which by DeMorgan becomes $\sim H_1 v \sim H_2 v \sim H_3$. In other words, perhaps the current hypothesis is correct, and one of the others is wrong. This analysis underlies what is called "saving the appearances", that is, holding fast to our hypothesis in spite of disconfirming evidence.

Suppose we are trying to discredit the flat earth hypothesis. As evidence, we say that if the earth is flat, then a ship sailing away from us, as it gets farther and farther away, should finally disappear all at once. But what we observe is that the ship disappears bit-by-bit, first the lower part, then the deck, and finally any tall structures. This observation tends to discredit the flat-earth hypothesis. However, a person who wished to defend the hypothesis could do so by denying that light travels in a straight line. This assumption, he could say, is what discredits his hypothesis, but in fact light travels in a curved fashion, thus giving the appearance that the ship is disappearing over a horizon.

What constitutes a confirming instance?

- Sheer accumulation of instances, except that one more instance is worth more in a small pool than in a large one.
- Instance variety: To prove that Community is Louisiana's favorite coffee, one more Cajun who likes it is not worth as much as a redneck from Winn Parish.
- Instances that not only tend to confirm the current hypothesis, but also tend to disconfirm a rival hypothesis.

Appendix

Answers to Odd-Numbered Exercises

Mental Calisthenics

Identifying problems

1. It is illegal to bury a living person.

3. 9 – the answer is given – it is a trick question, not a math question

5. The beggar is also a woman. They are sisters.

7. No. If his wife is a widow, it's because he is dead.

9. The trunk half full of dimes. A given quantity of nickels, such as 100, and half that quantity of dimes, such as 50, are equal in value. However, dimes are smaller than nickels, so more of them can fit into the same space. Half a trunk of dimes, then, is worth more than the same size trunk full of nickels.

11. The President. The VP's death does not remove the President from office.

13. Your mom.

15. It will fall faster through water at 60 degrees, because at 20 degrees the water will be frozen.

Math word problems

1. 3 – I will have either 3 of the same color, or 2 of one color, and one of the other color.

3. If the stream is shallow enough, they can wade. Otherwise, the 2 sons can cross together from Side A to Side B. Then one son can take the boat back across to his father on Side A. The father can cross alone to Side B. At that point, the son on Side B can take the boat back to Side A, and both sons can, again, cross to Side B. At that point, all 3 people (and the boat) will be on Side B.

5. March 22 & April 25 – Easter is defined as the Sunday following the first full moon of spring. The church calendar assumes that spring starts each year on March 21. If March 21 is a Saturday and also a full moon, Easter Sunday will be the next day. If the first full moon occurs a full month after March 21, it will fall on April 18. (21 + 28 days in a lunar month – 31 days in March = April 18.) If that is a Sunday, the next Sunday will be 7 days later = April 25.

Positions puzzles

3. Boys: Tom, Dick, and Harry
Girls: April, Carol, and Jane

Clues: Boys and girls must alternate. Tom not with Jane. Jane between Dick and Harry. April = position 1. Harry = position 6.

Seats: A T C D J H

Relationship puzzles

3. Girls = April, Betty, Carol
Majors = Literature, Biology, Engineering

Clues: Betty - not biology; Carol – not engineering, likes reading; April – not literature and not engineering

April = Biology; Betty = Engineering; Carol = Literature

5. The statement that the nutria finished two spaces ahead of the duck is ambiguous. It could mean that the nutria was two positions ahead of the duck or that there were two positions between the nutria and the duck. Since there are only four positions, the nutria must be in either first or second place. But we're told that the nutria was not second. So he must be first and the duck last. Note: there is not enough information to determine who finished second and third.

Matrix puzzles

1. We don't have enough information to complete the last matrix, but we need it to complete the other two.

States	Sen. Jones	Sen. Smith	Sen. Watson
AL	N	N	Y
FL	Y	N	N
TN	N	Y	N

Positions	Jones	Smith	Watson
Pilot	Y	N	N
Co-pilot	N	N	Y
Navigator	N	Y	N

Positions	AL	FL	TN
Pilot		N	
Co-pilot		N	
Navigator	N	Y	N

Logic and Language

Language forms and language functions

3. Declarative form; inspirational or ceremonial function, depending upon whether it's from the heart or perfunctory

5. Declarative form; informative function. Some people would say it is leaning toward the impressive category.

7. Poetic; beautiful – it lends a touch of elegance to the simple things of life.

9. The form is declarative, but Darwin is using a bit of sarcasm to make a point.

11. Declarative; information – some may say that it's also impressive, but there really isn't a better way to state that. You can use simpler language, but then you must use multiple sentences. This is as concise and straight-forward as it can be.

13. This seems to be both declarative and inspirational. There is a bit of poetic rhythm, also.

Ambiguity

3. Ambiguity on the word "sound". It can refer to an auditory sensation or the physicist's definition of a wave of a certain frequency.

5. The ambiguity is caused by an awkward grammatical structure. We don't know who slapped whom.

Vagueness

3. How much is excessive? Nothing precise is said.

5. Preferred by whom? The advertisers obviously want you to think it's preferred by high-class people, but it may actually be preferred by skid-row bums.

Analogies

3. The two games are both team sports, and apples and oranges are both fruits. But the two fruits are very different. As far as it goes, it is a good analogy.

5. This was obviously written before people had individual electronic devices. It may once have been an apt analogy, but no longer.

7. For those of us whose vision is not severely impaired, most of our images of the world are visual. We even dream primarily in pictures rather than sounds or smells.

9. This is an apt analogy for some businesses, but there are enough exceptions to limit its applicability.

11. This is obviously a politically biased comparison.

13. Well-written but obviously prejudiced.

15. Heisenberg seemed to be undermining the deterministic view that had been the prevailing assumption of physicists for many years, but it's probably not a fair comparison of Heisenberg himself.

Arguments

Christmas Carols

3. Angels from the Realms of Glory
5. It Came Upon a Midnight Clear
7. I'm Dreaming of a White Christmas
9. Away in a Manger
11. O Little Town of Bethlehem
13. O Come All Ye Faithful!
15. The Little Drummer Boy
17 Hark the Herald Angels Sing

Disagreements

1. Factual disagreement about the author of *The Old Man and the Sea.*

3. Interpretive disagreement about what makes for a good route, shortness or roadside beauty.

5. Evaluative disagreement about what constitutes a good time to buy a house.

Recognizing arguments

In the following exercises, P indicates premise and C indicates conclusion. There is only one premise unless otherwise stated.

1. P: "America doesn't work if you don't vote."
   C: "You need to vote."

3. P: "we are sinners all"
   C: "Forbear to judge."

5. P_1: "If the world were flat, the ship would disappear all at once when it reached the edge of the world."
   P_2: "The ship will disappear from sight gradually as it goes over the horizon."
   C: "the world is round rather than flat"

Arguments and explanations

1. It is an argument: it has a premise and a conclusion.

3. It is an explanation of why the police focus on the first 48 hours.

5. It is an explanation of the limited scope of general rules.

7. It is an argument to prove the need for nuclear power.

9. It is an explanation of the link between colonialism and the spread of Christianity.

Informal Logic

Fallacies of Ambiguity

1. Division: applying the characteristics of the group (traffic accidents) to a sub-group (traffic accident involving Model T Fords).

3. Division: applying the characteristics of two groups (cars and pickups) to individual members of both groups.

5. Composition: applying the characteristics of the individual monthly payments to the sum of all payments.

7. Equivocation on the word "character": a cartoon character versus a keyboard character.

Fallacies of Irrelevant Premises

1. Missing the point: the fact that a policy is being widely adopted doesn't show that it is a good policy. It could also be called *ad populum* – the bandwagon effect.

3. *Ad hominem*: an implicit put-down of anyone who doesn't share your point of view. It could also be called poisoning the well (see fallacies of unfair premises).

5. Another example of *ad hominem*.

7. This starts off as an appeal to pity and then turns into a threat.

9. "*Tu quoque* – two wrongs make a right."

Fallacies of Inadequate Premises

1. False cause (*post hoc*) – just because B follows A does not mean that A causes B.

3. Same as #1.

5. Circular argument.

7. Ditto.

9. Appeal to ignorance.

Fallacies of Unfair Premises

1. Slippery slope (domino theory).

3. Ditto.

5. This could be considered a persuasive definition, because Johnson is in effect redefining what is meant by the right of dissent.

7. This could be considered a false dilemma, since the person is taking an all-or-nothing attitude toward government benefits. It could also be called slippery slope.

9. This could be considered straw man or poisoning the well. The attempt is to prevent a fair discussion of the issues.

All Fallacies

1. *Tu quoque* – two wrongs make a right.

3. Composition – saying that just because you're only losing a little money each time, therefore you're not losing much money overall.

5. Circular argument: it says in effect that democracy is good because it's democracy.

7. Hasty generalization (converse accident): the generalization is based on a group that may not be representative of the whole population.

9. False cause (*post hoc*): saying that A caused B just because A preceded B.

11. Appeal to force. This is a blatant threat.

13. Amphiboly. The sentence structure is capable of 2 interpretations. The father meant that people take dogs there to race other dogs, but the son thought he meant that the people race the dogs.

15. Missing the point. Unfortunately, a politician's lack of qualifications doesn't prevent him from being elected.

17. Questionable analogy. The differences between the things being compared are more important that their similarities.

19. Division. Applying the characteristics of the whole (the country) to a part (an individual). This mistakenly assumes that America's wealth is evenly distributed.

21. Composition. This fallacy moves in the opposite direction of the preceding one, applying the characteristics of members of the group to the whole group.

23. *Tu quoque* – two wrongs make a right. The senator is trying to excuse his own sins by implying that all of us are sinners.

25. Slippery slope.

27. A humorous use of false cause (*non causa*).

29. It could be *ad hominem* or poisoning the well. The fallacy is that the speaker is being discredited without examining what he says.

Formal Logic

Standard Form

5. No people over 65 are people who are eligible to claim the earned income tax credit.

7. Some of my recipes are modifications of recipes that I have gotten from other people.

9. All times Harry has a good time are times when he has lots of money to spend. or No times when Harry does not have lots of money to spend are times he has a good time.

11. Some intelligent and sensitive physicians are not board-certified physicians.

13. Some people are not people who walk ten miles to school uphill in the snow.

15. Some people are people who have lived to regret a misspent youth.

17. Some things are not things that are both safe and exciting.

19. Some times are not times when good advice is appreciated. or Some people are not people who appreciate good advice.

21. No easy exercises are exercises that give me a headache.

23. All things of beauty are things that are joys forever.

Inference Along the Sides of the Square

1. RiT is the subcontrary of RoT.
   If RiT is true, RoT is undetermined.
   If RiT is false, RoT is true.

3. SaM is the super-altern of SiM.
   If SaM is true, SiM is true.
   If SaM is false, SiM is undetermined.

5. KiD is the sub-altern of KaD.
   If KiD is true, KaD is undetermined.
   If KiD is false, KaD is false.

7. PoO is the subcontrary of PiO.
   The truth values are the same as for #1.

Other Immediate Inference – Conversion

1. The converse of EeB is BeE. They are equivalent.
3. The converse of KiL is LiK. They are equivalent.
5. The converse of OoR is RoO. They are not equivalent.
7. The converse of LeV is VeL. They are equivalent.
9. The converse of SaF is FaS. They are not equivalent.

Other Immediate Inference – Obversion

1. The obverse of BiT is Bo$\overline{T}$.
3. The obverse of PaN is Pe$\overline{N}$.
5. The obverse of VaR is Ve$\overline{R}$.
7. The obverse of SiR is So$\overline{R}$.
9. The obverse of He$\overline{M}$ is HaM.

Sets of Exercises – first set

0. SaP is true	0. SaP is false
1. Se$\overline{P}$ obverts to SaP (the original) – true	1. false
3. $\overline{P}$a$\overline{S}$ ↔ SaP (contrapositive) – true	3. false
5. SoP contradictory of original – false	5. true
7. $\overline{P}$iS converts to Si$\overline{P}$ which obverts to SoP contradictory of original – false	7. true

Sets of Exercises – second set

0. SeP ↔ PeS (converse) – both are true	0. SeP ↔ PeS both are false
1. $\overline{P}$eS ↔ Se$\overline{P}$ ↔ SaP (converse and obverse) contrary of SeP – false	1. undetermined

3. $\overline{S}a\overline{P} \leftrightarrow PaS$ (contrapositive) contrary of PeS – false	3. undetermined
5. $\overline{S}e\overline{P}$ contrary of #3 – undetermined	5. undetermined
7. $\overline{P}iS \leftrightarrow Si\overline{P} \leftrightarrow SoP$ (converse and obverse) sub-altern of SeP – true	7. undetermined
9. SoP sub-altern of SeP – true	9. undetermined

<u>Sets of Exercises – third set</u>

0. $SiM \leftrightarrow MiS$ (converse) – both are true	0. SiM and MiS are false
1. SaM super altern of SiM – undetermined	1. false
3. $\overline{S}eM \leftrightarrow Me\overline{S} \leftrightarrow MaS$ (converse and obverse) super altern of MiS – undetermined	3. false
5. $Mo\overline{S} \leftrightarrow MiS$ (obverse)– true	5. false
7. $\overline{S}o\overline{M} \leftrightarrow MoS$ (contrapositive) – sub-contrary of MiS – undetermined	7. true
9. SeM contradictory of SiM – false	9. true
11. $\overline{S}oM$ sub-altern of #3 – undetermined	11. undetermined
13. $Se\overline{M} \leftrightarrow SaM$ (obverse) – super-altern of SiM – undetermined	13. false
15. $MeS \leftrightarrow SeM$ (converse) – contradictory of SiM – false	15. true
17. $\overline{M}e\overline{S} \leftrightarrow \overline{S}e\overline{M}$ (converse) – super-altern of #7 – undetermined	17. undetermined

19. $\overline{M}aS \leftrightarrow \overline{M}e\overline{S} \leftrightarrow \overline{S}e\overline{M}$ (obverse and converse) – super-altern of #7 – undetermined	19. undetermined
21. MoS – sub-contrary of MiS – undetermined	21. true
23. MiS – true	23. false
25. $\overline{M}a\overline{S} \leftrightarrow SaM$ (contrapositive) super-altern of SiM – undetermined	25. false
27. SoM sub-contrary of SiM – undetermined	27. true
29. $Mi\overline{S} \leftrightarrow MoS$ (obverse) sub-contrary of MiS – undetermined	29. true
31. $\overline{M}oS$ contradictory of #19 – undetermined	31. undetermined

<u>Sets of Exercises – fourth set</u>

0. $MoP \leftrightarrow Mi\overline{P} \leftrightarrow \overline{P}iM$ (obverse and converse) All of these are true.	0. $MoP \leftrightarrow Mi\overline{P} \leftrightarrow \overline{P}iM$ All of these are false.
1. $\overline{P}o\overline{M} \leftrightarrow MoP$ (contrapositive) – true	1. false
3. $Ma\overline{P} \leftrightarrow MeP$ super-altern of MoP – undetermined	3. false
5. MeP super-altern of MoP – undetermined	5. false
7. $\overline{P}aM$ super-altern of $\overline{P}iM$ – undetermined	7. false
9. $Po\overline{M} \leftrightarrow PiM \leftrightarrow MiP$ (obverse and converse) sub-contrary of MoP – undetermined	9. true
11. MiP sub-contrary of MoP – undetermined	11. true

13. Pa$\overline{\text{M}}$ ↔ PeM ↔ MeP (obverse and converse) super-altern of MoP – undetermined	13. false
15. $\overline{\text{P}}$i$\overline{\text{M}}$ sub-contrary of #1 – undetermined	15. true
17. $\overline{\text{P}}$iM ↔ Mi$\overline{\text{P}}$ ↔ MoP (converse and obverse) – true	17. false
19. $\overline{\text{M}}$o$\overline{\text{P}}$ ↔ PoM (contrapositive) no basis of comparison – undetermined	19. undetermined
21. $\overline{\text{M}}$iP ↔ Pi$\overline{\text{M}}$ ↔ PoM (converse and obverse) no basis of comparison – undetermined	21. undetermined
23. Me$\overline{\text{P}}$ ↔ MaP (obverse) contradictory of MoP – false	23. true
25. MaP contradictory of MoP – false	25. true
27. $\overline{\text{M}}$i$\overline{\text{P}}$ ↔ $\overline{\text{P}}$i$\overline{\text{M}}$ see #15 – undetermined	27. true

<u>Translation Exercises</u>

1. AoD – true	AoD – false
$\overline{\text{D}}$iA ↔ Ai$\overline{\text{D}}$ ↔ AoD (converse and obverse) – true	false
$\overline{\text{A}}$oD ↔ $\overline{\text{D}}$oA (contrapositive) – sub-contrary of previous – undetermined	true
DiA ↔ AiD (converse)	
sub-contrary of AoD – undetermined	true

3. Gi$\overline{V}$ ↔ GoV (obverse) – both are true	Gi$\overline{V}$ ↔ GoV – both are false
VeG ↔ GeV (converse) super-altern of GoV – undetermined	false
GaV contradictory of GoV – false	true
$\overline{V}$iG↔Gi$\overline{V}$ (converse) – true	false
5. $\overline{M}$a$\overline{I}$ ↔ IaM (contrapositive) – both are true	$\overline{M}$a$\overline{I}$ ↔ IaM – both are false
Ii$\overline{M}$ ↔ IoM (obverse) contradictory of IaM – false	true
MaI no basis of comparison – undetermined	underdetermined
Mi$\overline{I}$ ↔ MoI (obverse) no basis of comparison – undetermined	underdetermined

Extended Inference

Syllogisms

1. FaT
   FaI Drawing a particular conclusion from universal premises
   IiT

3. DeW
   JoD
   JiW

Exclusive premises or Drawing an affirmative conclusion from a negative premise

5. BeN
   BaF Illicit minor
   FeN

7. DaC
   LaD Valid
   LaC

Making arguments into syllogisms

1. MiR	MiR
$Je\overline{M} \leftrightarrow JaM$	JaM
JiR	JiR

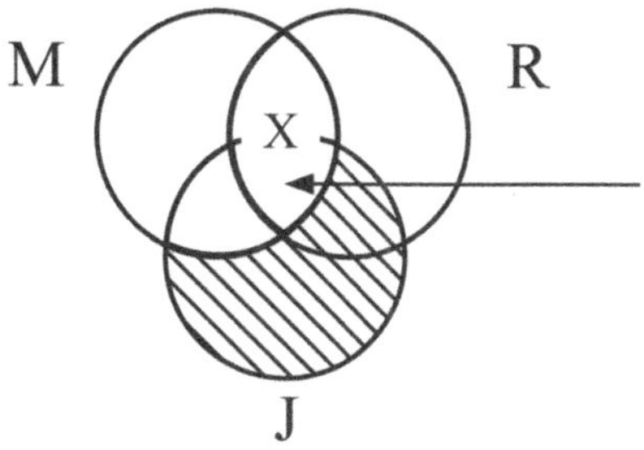

Undistributed middle term

3.	EaI	EaI
	Ia$\overline{S}$ ↔ IeS	<u>IeS</u>
	Sa$\overline{E}$ ↔ SeE	SeE
		Valid

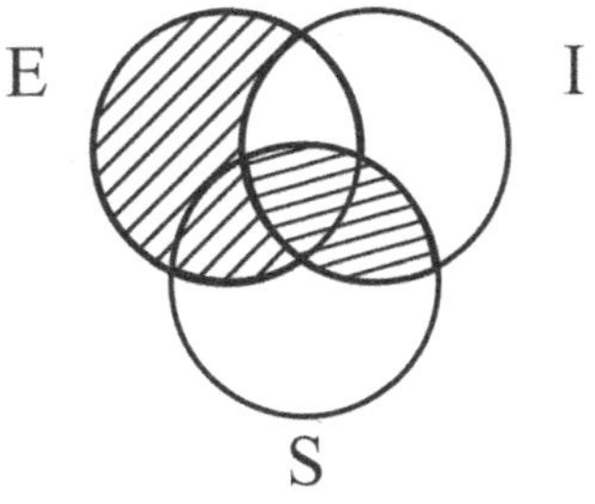

5. Using P for People who are included

$\overline{D}$a$\overline{P}$ ↔	PaD	
PeI	<u>PeI</u>	
IeD	IeD	Illicit Major

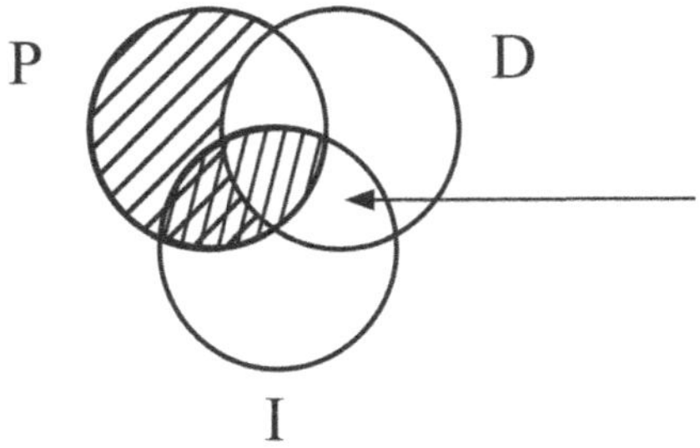

7. Using N for No less than 6 feet tall

BaS	SaN
$\overline{N}$eS ↔ Se$\overline{N}$ ↔ SaN	<u>BaS</u>
BaN	BaN
	Valid

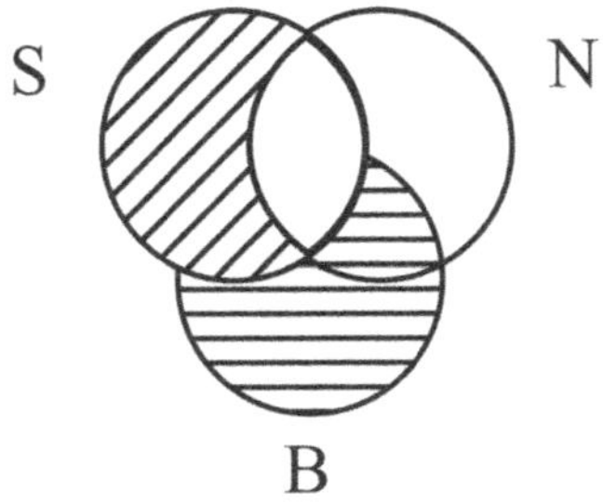

9. $\text{Na}\overline{\text{V}} \leftrightarrow \text{NeV}$

ViS	ViS
SoN	SoN
Valid	

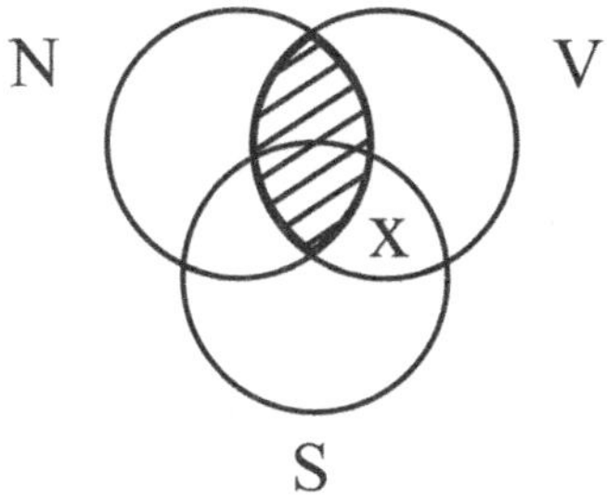

11. Using L for Lazy and U for Unlucky

$\text{Ra}\overline{\text{U}} \leftrightarrow \text{ReU}$
LiR
$\overline{\text{L}}\text{iU} \leftrightarrow \text{Ui}\overline{\text{L}} \leftrightarrow \text{UoL}$

LiR
ReU
UoL
Illicit major

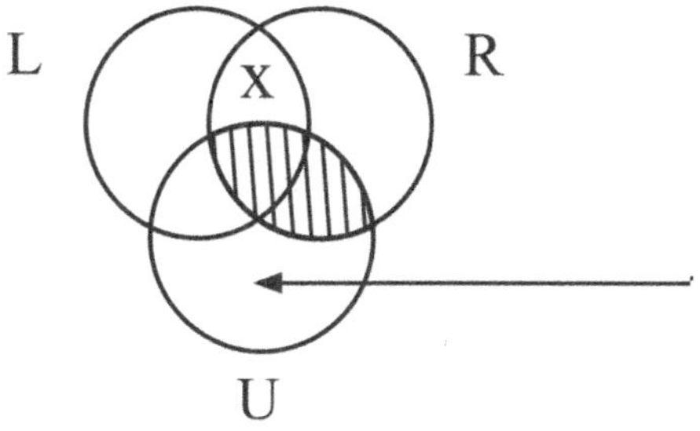

13. FeR
TaR
TeF
Valid

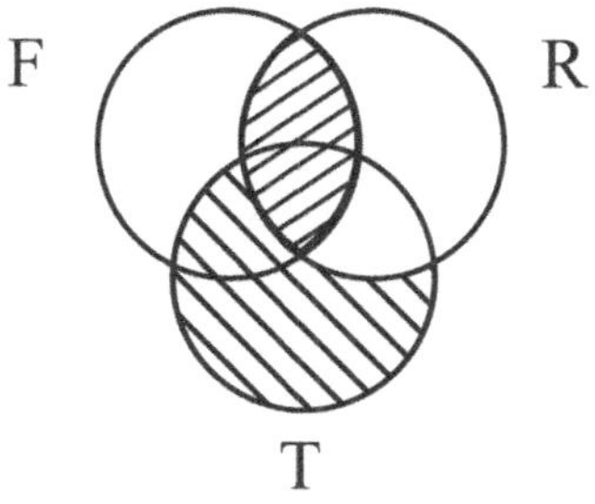

15. $\overline{R}eC \leftrightarrow Ce\overline{R} \leftrightarrow$ CaR
VaC VaC
VaR VaR
Valid

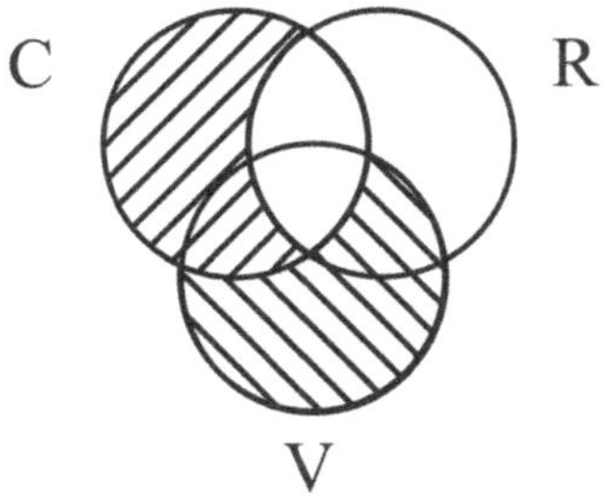

Enthymemes

The supplied premise or conclusion will be indicated with an arrow like this →

3. → BaC All blessed people are called children of God.
PaB
PaC

5. → Wa21 People who can buy whiskey are 21.
He21
HeW

7 EeC
TaE
→ TeC This argument is not complete.

9. EaR
YaE
→ YaR You must respond to your critics.

11. TaD
→ PoD Some people don't pass difficult courses.
PoT

13. PaM Note: P stands for people who passed this exam.
YeM
→ YeP You didn't pass this exam.

15. → WaR Whenever I wash my car it rains.
<u>TaW</u>
TaR

17. DeC
<u>DiB</u>
→ BoC Some beagles are not cats.

<u>Sorites</u>

1. EaS
SaL
<u>LaU</u>
EaU

Constituent syllogisms:

SaL	LaU
<u>EaS</u>	<u>EaL</u>
EaL	EaU

3. MaS
SaP
<u>BeM</u>
BeP
There's no need going any further with this one. We can see that the major term (P) is distributed in the conclusion but not in the premise (Illicit Major).

5. MaE
PeE
OiW
<u>OaM</u>
PoW
Same problem as #3. The major term (W) is distributed in the conclusion but not in the premise.

7. This is so involved that we need to define our terms before trying to do anything with it.

S: anything that promotes skin cancer
D: anything that causes death
O: whatever preserves the ozone layer
R: whatever allows the release of chlorofluorocarbons
U: whatever increases ultraviolet radiation
M: packaging materials
L: things that are legal

SaD	Constituent syllogisms:
Oa$\overline{R}$ ↔ Ra$\overline{O}$	SaD
$\overline{S}$eU ↔ Ue$\overline{S}$ ↔ UaS	$\underline{\text{DeL}}$
$\overline{O}$aU	LeS
MiR	$\underline{\text{UaS}}$
DeL	UeL
Mi$\overline{L}$ ↔ MoL	$\underline{\overline{\text{O}}\text{aU}}$
	$\overline{O}$eL
	$\underline{\text{Ra}\overline{\text{O}}}$
	ReL
	$\underline{\text{MiR}}$
	MoL

Enthymemic Sorites

1\. FaD — Constituent syllogisms:

FaD	Constituent syllogisms:		
DaM	DaM	MaA	AaL
MaA	$\underline{\text{FaD}}$	$\underline{\text{FaM}}$	$\underline{\text{FaA}}$
AaL	FaM	FaA	FaL
			Fido is a living thing.

3\. Using L for Logic and E for large Enrollment.

LaD	Constituent syllogisms:	
DoE ↔ $\overline{E}$o$\overline{D}$ ↔ $\overline{E}$iD	$\overline{E}$iD	CiD
$\overline{E}$aC	$\underline{\overline{\text{E}}\text{aC}}$	$\underline{\text{LaD}}$
	CiD	LiC

This looked promising, but the last syllogism gives us Undistributed Middle

Mathematical Logic

<u>Truth Tables</u>

1. A OR (A AND B)

A	B	A AND B	A OR (A AND B)
T	T	T	T
T	F	F	T
F	T	F	F
F	F	F	F

3. A OR (B AND C)

A	B	C	B AND C	A OR (B AND C)
T	T	T	T	T
T	T	F	F	T
T	F	T	F	T
T	F	F	F	T
F	T	T	T	T
F	T	F	F	F
F	F	T	F	F
F	F	F	F	F

5. NOT [A AND NOT (A OR NOT B)]

1	2	3	4	5	6	7
A	B	NOT B	A OR NOT B	NOT (A OR NOT B)	A AND 5	NOT 6
T	T	F	T	F	F	T
T	F	T	T	F	F	T
F	T	F	F	T	F	T
F	F	T	T	F	F	T

Mathematical Symbolism

1. A•(AvB)

A	B	AvB	A•(AvB)
T	T	T	T
T	F	T	T
F	T	T	F
F	F	F	F

3. [(A → B)•A] → B

A	B	A→B	(A→B)•A	[(A→B•A]→B
T	T	T	T	T
T	F	F	F	T
F	T	T	F	T
F	F	T	F	T

5. ~(A → B) • (~BvC)

A	B	C	~B	~BvC	A→B	~(A→B)	ALL
T	T	T	F	T	T	F	F
T	T	F	F	F	T	F	F
T	F	T	T	T	F	T	T
T	F	F	T	T	F	T	T
F	T	T	F	T	T	F	F
F	T	F	F	F	T	F	F
F	F	T	T	T	T	F	F
F	F	F	T	T	T	F	F

Symbolic Translation Exercises

1. (S•~T) → R
3. (S•T) → ~R
5. ~P → ~R
7. (GvF) → I
9. (FvI) → G

More Symbolic Translation Exercises

1. "No easy exercise ever gives me a headache." E: these are easy exercises. H: these exercises give me a headache. (e) E →~H

3. "If you are not good in math, then you won't do well in logic and you won't be able to satisfy the college's critical reasoning requirement." G: You are good in math. L: You do well in logic. S: You will be able to satisfy the college's critical reasoning requirement. (b) ~G → (~L•~S)

5. "Germany will mobilize its troops and either France will declare an emergency or Italy will appeal to the European Union." G: Germany will mobilize its troops. F: France will declare an emergency. I: Italy will appeal to the European Union.. (c) G • (FvI)

Still More Symbolic Translation Exercises

1. F•(S•L)
3. ~(C•D)
5. S →~H
7. E →~C
9. Cv~C
11. [(~U →~R) • (~R →~U)]•N

Computer Logic

<u>Circuit Design</u>

1. ~O•(FvB)

B	F	O	~O	FvB	~O•(FvB)
T	T	T	F	T	F
T	T	F	T	T	T
T	F	T	F	T	F
T	F	F	T	T	T
F	T	T	F	T	F
F	T	F	T	T	T
F	F	T	F	F	F
F	F	F	T	F	F

3.

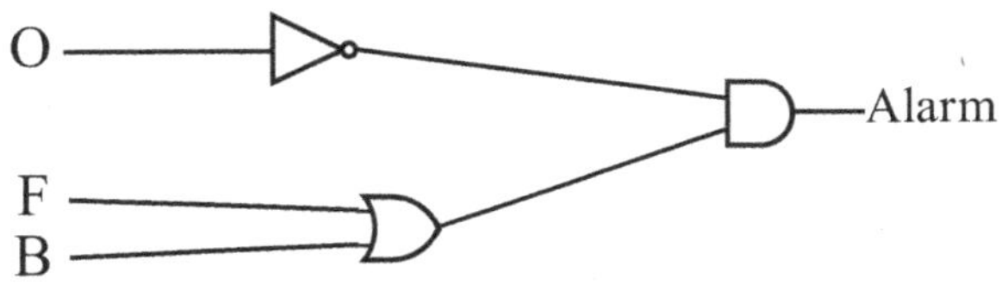

5. (~A•B)v~C

7.

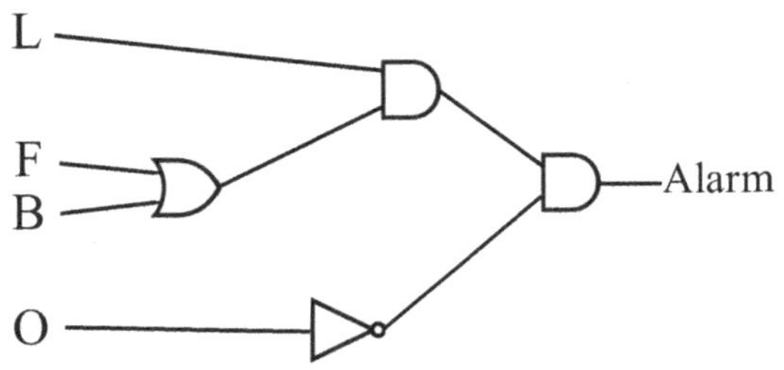

Decision Structures

1.

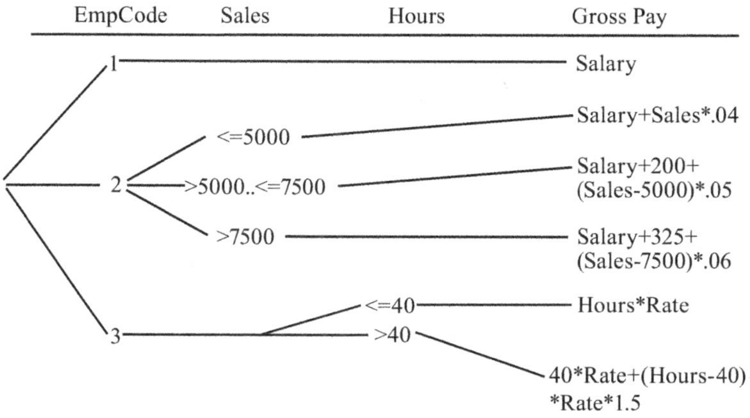

Probability

The Probability Calculus

1. (a) 4/52*4/52*4/52 = .00045
   (b) 4/52*3/51*2/50 = .000181

3. (a) p(D•D) = 7/28*6/27 = .056
   (b) 1- p(~D•~D) = 1-(21/28*20/27) = .44

5. (a) p(BA•BA) = 2/52*1/51 = .0008
   (b) p(~BA•~BA) = 50/52*49/51 = .9238

7. (a) p(A•A•A•A) = 4/52*3/51*2/50*1/49 = .0000036
   (b) p(AvAvAvA) = 1-p(~A•~A•~A•~A) =
   1-(48/52*47/51*46/50*45/49) = .2812
   (c) p(A•~A•~A•~A) v(~A•A•~A•~A)v(~A•~A•A•~A)v
   (~A•~A•~A•A)= 4/52*48/51*47/50*46/49*4 = .2506

9. (a) 1-p(~S•~S•~S) = 1-(3/4*3/4*3/4) = .5781
   or p[(S•~S•~S)v(~S•S•~S)v(~S•~S•S)v(S•S•~S)v(~S•S•S)
   v(S•~S•S)v(S•S•S)] = (1/4*3/4*3/4)+(3/4*1/4*3/4)+
   (3/4*3/4*1/4)+(1/4*1/4*3/4)+(3/4*1/4*1/4)+(1/4*3/4*1/4)
   +(1/4*1/4*1/4) = .5781
   (b) 1- p(~S•~S•~S) = 1-(39/52*38/51*37/50) = .4135

11. The odds of drawing another Jack when there's only one remaining are p[(J•~J)v(~J•J)] = (1/47*45/46)+(45/47*1/46) = .0425

The odds of drawing a pair when there are 12 pair remaining are p[(A•A)v(K•K) etc.] = 12*(4/47*3/46) = .0666

So the odds of drawing a Jack or a pair of something else would be the chances of doing either of them less the chances of doing them both: .0425+.0666-(.0425*.0666) = .1063 – not enough to justify a big bet!

Deductive Inference

Statements and Statement Forms

3. This is a substitution instance of a, d, h, and k.
5. This is a substitution instance of a and d.
7. This is a substitution instance of a and e.
9. This is a substitution instance of a and b.
11 This is a substitution instance of a and b.
13. This is a substitution instance of a and b.
15. This is a substitution instance of a, d, and k.
17. This is a substitution instance of a and d.

<u>Tautologies, contradictions, and contingent statements</u>

1. D →D Tauatologous.

D	D→D
T	T
F	T

3. M → (Lv~L) Tauatologous

M	L	~L	Lv~L	M→(Lv~L)
T	T	F	T	T
T	F	T	T	T
F	T	F	T	T
F	F	T	T	T

5. F→(T•~T) Contingent.

F	T	~T	T•~T	F→(T•~T)
T	T	F	F	F
T	F	T	F	F
F	T	F	F	T
F	F	T	F	T

7. V → E Contingent.

E	V	V→E
T	T	T
T	F	T
F	T	F
F	F	T

9. W → O Contigent.

O	W	W→O
T	T	T
T	F	T
F	T	F
F	F	T

11. [p → (p → q)] → q Contingent.

p	q	p→q	p→(p→q)	[p→(p→q)]→q
T	T	T	T	T
T	F	F	F	T
F	T	T	T	T
F	F	T	T	F

.

13. p → [~p → (qv~q)] Tautologous.

p	q	~p	~q	qv~q	~p→(qv~q)	p→[~p→(qv~q)]
T	T	F	F	T	T	T
T	F	F	T	T	T	T
F	T	T	F	T	T	T
F	F	T	T	T	T	T

15. p → [p → (q•~q)]   Contingent.

p	q	~q	q•~q	p→(q•~q)	p→[p→(q•~q)]
T	T	F	F	F	F
T	F	T	F	F	F
F	T	F	F	T	T
F	F	T	F	T	T

17. (p → q) ↔ (~p →~q)   Contingent.

p	q	~p	~q	p→q	~p→~q	(p→q)↔(~p→~q)
T	T	F	F	T	T	T
T	F	F	T	F	T	F
F	T	T	F	T	F	F
F	F	T	T	T	T	T

Equivalence Exercises

In these exercises, an arrow (←) will be used to mark the rows in which the two expressions are not equivalent, if there are any.

1. Is p → q equivalent to ~p →~q   Not equivalent

p	q	~p	~q	p→q	~p→~q	
T	T	F	F	T	T	
T	F	F	T	F	T	←
F	T	T	F	T	F	←
F	F	T	T	T	T	

3. Is (p → q) → r equivalent to (q → p) → r? Not equivalent.

p	q	r	p→q	(p→q)→r	q→p	(q→p)→r	
T	T	T	T	T	T	T	
T	T	F	T	F	T	F	
T	F	T	F	T	T	T	
T	F	F	F	T	T	F	←
F	T	T	T	T	F	T	
F	T	F	T	F	F	T	←
F	F	T	T	T	T	T	
F	F	F	T	F	T	F	

5. Is p equivalent to pv(p•q)? Yes, they are equivalent.

p	q	p•q	pv(p•q)
T	T	T	T
T	F	F	T
F	T	F	F
F	F	F	F

7. Is p equivalent to p•(q → p)? Yes, they are equivalent.

p	q	q→p	p•(q→p)
T	T	T	T
T	F	T	T
F	T	F	F
F	F	T	F

9. Is p•(qvr) equivalent to (pvq)•(pvr)? Not equivalent.

p	q	r	qvr	p•(qvr)	pvq	pvr	(pvq)•(pvr)	
T	T	T	T	T	T	T	T	
T	T	F	T	T	T	T	T	
T	F	T	T	T	T	T	T	
T	F	F	F	F	T	T	T	←
F	T	T	T	F	T	T	T	←
F	T	F	T	F	T	F	F	
F	F	T	T	F	F	T	F	
F	F	F	F	F	F	F	F	
Columns being compared.				↑			↑	

11. Is (p•q) → r equivalent to p → (q → r)? Equivalent.

p	q	r	p•q	(p•q)→r	q→r	p→(q→r)
T	T	T	T	T	T	T
T	T	F	T	F	F	F
T	F	T	F	T	T	T
T	F	F	F	T	T	T
F	T	T	F	T	T	T
F	T	F	F	T	F	T
F	F	T	F	T	T	T
F	F	F	F	T	T	T
Columns being compared.				↑		↑

Implication Exercises

1. Does (p → q)•~q logically imply ~p? Yes, since the statement of the implication is a tautology.

1	2	3	4	5	6	7
p	q	~p	~q	p→q	(p→q)•~q	6→3
T	T	F	F	T	F	T
T	F	F	T	F	F	T
F	T	T	F	T	F	T
F	F	T	T	T	T	T

3. Does (p → q)•(q → r) logically imply p → r? Yes.

1	2	3	4	5	6	7	8
p	q	r	p→q	q→r	(p→q)•(q→r)	p→r	6→7
T	T	T	T	T	T	T	T
T	T	F	T	F	F	F	T
T	F	T	F	T	F	T	T
T	F	F	F	T	F	F	T
F	T	T	T	T	T	T	T
F	T	F	T	F	F	T	T
F	F	T	T	T	T	T	T
F	F	F	T	T	T	T	T

5. Does (pvq)•~p logically imply q? Yes.

1	2	3	4	5	6
p	q	~p	pvq	(pvq)•~p	5→2
T	T	F	T	F	T
T	F	F	T	F	T
F	T	T	T	T	T
F	F	T	F	F	T

7. Does p•q logically imply p? Yes.

p	q	p•q	(p•q)→p
T	T	T	T
T	F	F	T
F	T	F	T
F	F	F	T

9. Does (p•q) → r logically imply p → (q→r)? Yes.

1	2	3	4	5	6	7	8
p	q	r	p•q	(p•q)→r	q→r	p→(q→r)	5→7
T	T	T	T	T	T	T	T
T	T	F	T	F	F	F	T
T	F	T	F	T	T	T	T
T	F	F	F	T	T	T	T
F	T	T	F	T	T	T	T
F	T	F	F	T	F	T	T
F	F	T	F	T	T	T	T
F	F	F	F	T	T	T	T

11. Does [(p•q) → (r•s)]•(pvr) imply qvs? No.

1	2	3	4	5	6	7	8	9	10		
p	q	r	s	p•q	r•s	5→6	pvr	7•8	qvs	9→10	
T	T	T	T	T	T	T	T	T	T	T	
T	T	T	F	T	F	F	T	F	T	T	
T	T	F	T	T	F	F	T	F	T	T	
T	T	F	F	T	F	F	T	F	T	T	
T	F	T	T	F	T	T	T	T	T	T	
T	F	T	F	F	F	T	T	T	F	F	←
T	F	F	T	F	F	T	T	T	T	T	
T	F	F	F	F	F	T	T	T	F	F	←
F	T	T	T	F	T	T	T	T	T	T	

(continued from previous page)

p	q	r	s	p•q	r•s	5→6	pvr	7•8	qvs	9→10	
F	T	T	F	F	F	T	T	T	T	T	
F	T	F	T	F	F	T	F	F	T	T	
F	T	F	F	F	F	T	F	F	T	T	
F	F	T	T	F	T	T	T	T	T	T	
F	F	T	F	F	F	T	T	T	F	F	←
F	F	F	T	F	F	T	F	F	T	T	
F	F	F	F	F	F	T	F	F	F	T	

<u>Assigning Truth Values</u>

1. L• ~S
T• ~F
T

3. M• ~L
T• ~F
T

5. ~L → M
~F → T
T

7. ~(L•W)
~(T•F)
T

9. $(L_2 \cdot W_2) \rightarrow (L \cdot W)$
(T•T) → (T•F)
T → F
F

11. $(N_2 \cdot T) \rightarrow U$
(F•T) → T
F → T
T

13. $(N \cdot {\sim} N_2) \cdot U$
(T•~F)•T
(T•T)•T
T

Proving Invalidity

An arrow (←) will be used to mark rows in which the premises are true and the conclusion is false.

1. p → q ∴~p →~q

				P	C	
p	q	~p	~q	p→q	~p→~q	
T	T	F	F	T	T	
T	F	F	T	F	T	
F	T	T	F	T	F	←
F	F	T	T	T	T	

3. p → q, p → r ∴ qvr

			P_1	P_2	C	
p	q	r	p→q	p→r	qvr	
T	T	T	T	T	T	
T	T	F	T	F	T	
T	F	T	F	T	T	
T	F	F	F	F	F	
F	T	T	T	T	T	
F	T	F	T	T	T	
F	F	T	T	T	T	
F	F	F	T	T	F	←

5. pvq, p ∴ q

P_2	C	P_1	
p	q	pvq	
T	T	T	
T	F	T	←
F	T	T	
F	F	F	

<u>Proving Invalidity</u>

1. p→q ∴~p•~q
   F→T ~F•~T
   T F

3. p → q, p → r ∴ qvr
   F → F, F → F FvF
   T T F

5. pvq, p ∴ ~q
   TvT T ~T
   T T F

<u>Proving Validity (Re-visited)</u>

1. B →~T, ~T ∴ B
   F →~F ~F F
   T T F
   Invalid

3. W → L. W ∴ L
   F → F T F
   T T F
   Inconsistent assignment of truth values to W, therefore valid.

5. ~(S•C), ~(C•W), W ∴ ~(SvC)
   ~(T•F) ~(F• T) T ~(TvF)
   T T T F
   Invalid

7. A → B, B →~C ∴ A →~C
   T → T F →~T T →~T
   T T F
   Inconsistent assignment of truth values to B, therefore valid.

9. A → B, C → D, AvD ∴ BvC
   F → F F → T FvT FvF
   T T T F
   Invalid

11. Iv~J, ~(~K•L), ~(~I•~L) ∴ ~J → K
Tv~F ~(~F•F), ~(~T•~F) ∴ ~F → F
T T T F
Invalid

13. A•B, (AvC) → D ∴ A•D
T•T, (FvF) → F ∴ T•F
T T F
Inconsistent assignment of truth values to A, therefore valid.

15. I → J, J → K, L → M, IvL ∴ KvM
F → F, F → F, F → F, TvF ∴ FvF
T T T T F
Inconsistent assignment of truth values to I, therefore valid.

17. A ↔ (BvC), B ↔ (CvA), C ↔ (AvB), ~A ∴ BvC
F ↔ (FvF), F ↔ (FvF), F ↔ (FvF), ~F ∴ FvF
T T T T F
Invalid

Formal Proof of Validity

Arguments and Argument Forms

1. H.S. 3. Simp. 5. Add.

Formal Proof

(1)

(1)	1. A →B	
	2. B →~C	/∴ A →~C
	3. A →~C	1, 2 H.S.

(3)

(3)	1. A•B	
	2. (AvC) → D	/∴ A•D
	3.A	1 Simp.
	4. AvC	3 Add.
	5. D	2, 4 M.P.
	6. A•D	3, 5 Conj.

(5)	1. Fv(GvH)	
	2. (G → I) • (H → J)	
	3. (IvJ) → (FvH)	
	4. ~F	/∴ H
	5. G → I	2 Simp.
	6. H → J	2 Simp.
	7. GvH	1, 4 D.S.
	8. IvJ	5, 6, 7 C.D.
	9. FvH	3, 8 M.P.
	10. H	9, 4 D.S.

Rules of Replacement

1. D.N. 3. Exp. 5. DeM. 7. Equiv. 9. DeM.

Constructing proofs

(1)	1. A → B	
	2. C →~B	/∴ A →~C
	3. ~~B →~C	2 Contra.
	4. B →~C	3 D.N.
	5. A →~C	1,4 H.S.

(3)	1. (Qv~R)vS		
	2. ~Qv(R•~Q)	/∴ R → S	
	3. (~QvR) • (~Qv~Q)		2 Dist.
	4. ~Qv~Q		3 Simp.
	5. ~Q		4 Taut.
	6. Qv(~RvS)		1 Assoc.
	7. ~RvS		6, 5 D.S.
	8. R → S		7 Impl.

(5)	1. A → B		
	2. B → C		
	3. C → A		
	4. A →~C	/∴~A•~C	
	5. A → C		1, 2 H.S.
	6. (A → C)•C → A)		5, 3 Conj.
	7. A ↔ C		6 Equiv.
	8. (A•C) v (~A•~C)		7 Equiv.
	9. ~Av~C		5 Impl.
	10. ~(A•C)		9 DeM.
	11. ~A•~C		8, 10 D.S.

<u>Exercises, Part 1</u>

(1)	1 B → C	
	2. B	/∴B•C
	3. C	1, 2 M.P.
	4. B•C	2, 3 Conj.

(3)	1. ~C	
	2. A → C	
	3. AvB	/∴BvD
	4. ~A	2, 1 M.T
	5. B	3, 4 D.S.
	6. BvD	5, Add.

(5)	1. A	
	2. C	/∴ (AvB) • (CvD)
	3. AvB	1, Add.
	4. CvD	2, Add.
	5. (AvB) • (CvD)	3, 4 Conj.

(7)

1. A → B	
2. ~B	
3. (~A v ~C) → (D•E)	/∴ EvF
4. ~A	1, 2 M.T.
5. ~Av~C	4 Add.
6. D•E	3, 5 M.P.
7. E	6 Simp.
8. EvF	7 Add.

<u>Exercises, Part 2</u>

(1)

1. ~Av(B•C)	/∴ A → C
2. (~AvB)•(~AvC)	1 Dist
3. ~AvC	2 Simp
4. A → C	3 Impl.

(3)

1. (NvF) → (C•D)	
2. D → V	
3. V → I	
4. I → A	
5. A →~C	/∴ ~F
6. D → I	2, 3 H.S.
7. D → A	6, 4 H.S.
8. D →~C	7, 5 H.S.
9. ~Dv~C	8 Impl.
10. ~Cv~D	9 Comm.
11. ~(C•D)	10 DeM.
12. ~(NvF)	1, 11 M.T.
13. ~N •~F	12 DeM.
14. ~F	13 Simp.

(5)

1. A →~B	
2. AvC	
3. B•D	/∴ C
4. B	3 Simp.
5. ~~B	4 D.N.
6 ~A	1, 5 M.T.
7. C	2, 6 D.S.

(7)	1. [Av (B•C)]→(D•E)	
	2. ~E	/∴ ~A
	3. ~Ev~D	2 Add.
	4. ~Dv~E	3 Comm.
	5. ~(D•E)	4 DeM.
	6. ~[Av(B•C)]	1, 5 M.T.
	7. ~A•~(B•C)	6 DeM.
	8. ~A	7 Simp.

(9)	1. (O•T) → (S → M)	
	2. R →~M	
	3. T•R	
	4. O•S	/∴ V
	5. O	4 Simp.
	6. T	3 Simp.
	7. O•T	5, 6 Conj.
	8. S → M	1, 7 M.P.
	9. S	4, Simp.
	10. M	8, 9 M.P.
	11. ~~M	10 D.N.
	12.~R	2, 11 M.T.
	13.~RvV	12 Add.
	14. R	3 Simp.
	15. ~~R	14 D.N.
	16. V	13, 15 D.S.

(11)	1. (Dv~H) → [Rv(S•~T)]	
	2. L	
	3. (LvM) → (D•E)	/∴ T → R
	4. LvM	2 Add.
	5. D•E	3, 4 M.P.
	6. D	5, Simp.
	7. Dv~H	6 Add.
	8. Rv(S•~T)	1, 7 M.P.
	9. (RvS)•(Rv~T)	8 Dist.
	10. Rv~T	9 Simp.
	11. ~TvR	10 Comm.
	12. T→R	11 Impl.

Inductive Inference

Argument by Analogy

1. (a) More probable because of more instances.

(b) Mixed – it depends on his expectations. If he's expecting 10% appreciation, he may be setting himself up for disappointment. But if he's just expecting a 'pretty good' return, then he's more likely to get it.

(c) More probable – less competition.

(d) Neutral – probably makes no difference.

(e) Less probable – start-ups are very unpredictable.

3. (a) More probable, especially since Boudreaux likes to go fishing even if he doesn't catch many fish.

(b) Less probable – there's a big difference between not catching many and not catching any.

(c) Much more probable – a bad fishing day is made better by cold beer.

(d) Probably won't make any difference in his enjoyment.

Refutation by Logical Analogy

1. Argument being refuted: "cameras in the courthouse make the system look bad". Refuting analogy: press coverage of the Vietnam war made the U.S. involvement in the war look bad. Good analogy as far as it goes, but there are important differences. In some court cases, a lot is happening behind the scenes, such as plea bargaining.

3. Argument being refuted: "new highways bring induced traffic". Refuting analogy: building new grocery stores doesn't cause check-out lines to be longer. What is happening in both cases is a transfer of the traffic, not an increase in it.

Mill's Methods

1. She is using the Method of Concomitant Variation – the more white paint she adds, the less red the paint is.

3. I could give up wine for a while and see if it made a difference. In fact, I did just that – I gave up all alcohol for Lent, and it wasn't long before I was popping the antacid tablets on an almost daily basis.

5. This is obviously the Joint Method of Agreement and Difference.

7. The Method of Difference – I get the dog's weight by subtracting my weight from our combined weight.

9. If we can find reliable records going back far enough, we could find out what average temperatures were in pre-industrial times or when industry was not as wide-spread as it is today.

Made in the USA
Middletown, DE
04 November 2026

36544191R00136